D1067879

Moshava, Kibbutz, and Moshav

Patterns of Jewish Rural Settlement and Development in Palestine

Aerial view of Nahalal—1946. At the heart of the circle are located the various public buildings and the houses of artisans and other settlers not engaged in agriculture. The area that encircles this core is divided into wedge-shaped plots. Each settler's plot is attached to his house.

MOSHAVA, KIBBUTZ, AND MOSHAV

Patterns of Jewish Rural Settlement and Development in Palestine

D. WEINTRAUB, M. LISSAK, and Y. AZMON

Foreword by S. N. EISENSTADT

Cornell University Press

ITHACA AND LONDON

First published 1969

Standard Book Number: 8014-0520-3

Library of Congress Catalog Card Number: 69-18362

PRINTED IN THE UNITED STATES OF AMERICA
BY KINGSPORT PRESS, INC.

Foreword

This book is a comparative analysis of the three major types of agricultural settlements that have been developed in the Jewish community in Palestine and in Israel: the *moshava*, the first type of settlement to develop, based on individual, family-owned property; the *kibbutz*, a communal system; and the *moshav*, a cooperative type. As such, this work constitutes an important new approach to the study of Israeli society, since most previous analyses of Jewish settlement have tended to stress only the second or third type.

Moreover, the comparison of all the major settlement types is done within a general analytical framework. It does not treat each type as a group of single, autonomous villages, tied together only by common customs, background, or ideology; it shows in each case—especially in the kibbutz and the moshav—how individual villages developed from within broader socio-political and ideological movements that have established country-wide economic and political organizations. It analyzes how each village functions within these movements and demonstrates that its economic, political, and social patterns and activities cannot be fully understood without studying the movement to which it belonged. It also outlines the relation of the individual movements to the broader, general institutional framework and centers of the Yishuv, and later of the State of Israel.

The original tie between settlers and their movements was basically a motivational and ideological one. The original type of settlement which the settlers joined was mostly influenced, as the analysis presented in the book fully shows, by these motivational and ideological orientations and attitudes toward agriculture and by the settlers' perception of the place of such settlements and of agriculture in the general process of building up the new Jewish national community in Palestine and in Israel.

But once these settlements and movements were established, they generated their own institutional impetus, including also the more complex economic, political, and social relations within the broader societal framework. These relations regulated the flow of resources between the settlements and the wider social setting and greatly influenced the ways in which these various settlements adapted themselves to their environment—or to put it more accurately, the ways in which the movements attempted to shape their environment according to their own ideological orientations and institutional needs.

The importance of this analysis, however, goes far beyond the details of the Israeli scene, to the understanding of which it provides a very important contribution. Its analytical framework is also of great significance for the analysis of processes of agricultural change and development and of modernization of the rural sector in general. This can be seen from several points of view.

First, as this study shows, not just the internal structure, division of labor, and levels of resources of the rural sector and its components are important for an understanding of the processes of modernization. A full understanding of the ways in which these structures and resources can be utilized or modified or can impede development and modernization can be attained only if the various structural and symbolic links of this sector to the broader society, to its institutional frameworks and centers, also are analyzed.

Second, as the analysis indicates, both the internal structure of the agricultural sector and the nature of its links to the macrosocietal frameworks and centers differ not only among societies but also within the confines of one society: under different circumstances and conditions different parts of the agricultural sector may become predominant in influencing the sector's development.

Third, as this study sharply demonstrates, rural modernization and economic development of the agricultural sector need not necessarily always be the same, or at least they must be analytically distinguished. Modernization of the rural sector is mostly a sociological and political process characterized by growing social differentiation, political participation, and the like, while the development of the agricultural sector is by definition an economic process. Once this distinction has been made, the relations among these variables or aspects of development (as well as with another variable, agrarian reform, until now irrelevant in Israel) have to be studied. Further questions must be asked: Under what conditions does initial social and/or political modernization of the rural sector help or impede the economic development of the agricultural sector and its integration into the emerging modern economy? Conversely, under what conditions does an initial strong impetus in agricultural development help in the ultimate social and political modernization of the rural sector? Moreover, there is the question of how such combinations of these components, variables, or aspects of the rural sector may also affect the broader society's capacity for modernization and growth. The analysis presented here also indicates that the viability or success of any given type of modern agricultural sector, of any given institutional or organizational pattern within it, does not assure the sector's ability for continuous growth. For several of the major external circumstances may change in time, thus necessarily affecting the rural or agricultural sector itself. And these institutional settings and frameworks themselves, through their very success and institutionalization, may create ideological and institutional rigidities and vested interests that may not only prevent adjustments to new situations and problems but may even prevent them from being perceived as such.

One of the most fascinating problems in the study of modernization in general is, indeed, the analysis of the conditions under which any modern institutional setting, initially successful in creating a new complex and adaptable organization, may be able to overcome such rigidities and assure some measure of ability to deal continuously with new problems. This question is of special importance in the study of the agricultural sector, as this sector is doomed in any process of modernization to become less important.

The material in this book provides many fascinating illustrations and analyses of all these problems. The cases presented here are of special value, because several distinct phases in the development of each of the settlements are analyzed. The first phase was the initial establishment through the ideological impetus of the different Zionist groups. The second stage was the crystallization of the institutional settings of the settlements and of the movements in the period of the British mandate. Last—and perhaps most important in considering the problems outlined here—comes the phase of absorption of new immigrants from traditional settings, who came with the establishment of the State of Israel and who were absorbed and "modernized" with different degrees of success within the framework of these various settlements.

In each of these phases it was the interplay between the motivation and background of the settlers, the concrete institutional settings of the villages and movements, and the broader societal setting which created the different patterns—and problems—of development and modernization. Needless to say, the specific constellation of these various components is unique to Israel, as is the case in any single situation. But the special merit of the book lies first in the fact that it shows us how such specific analysis can be understood in terms of such broader analytic categories and how these categories can be applied to concrete situations. Second, it shows us that such application and comparison can and should be made both among societies and within them, and can lead to the reformulation of problems and to fruitful new hypotheses which can then again be tested on a broader comparative canvas.

S. N. Eisenstadt

The Eliezer Kaplan School of Economics & Social Science
The Hebrew University, Jerusalem, Israel
February 1968

Acknowledgments

The authors take great pleasure in acknowledging their debt of gratitude for support and help:

The United States Department of Agriculture, under whose grant the study was undertaken (Grant No. FG–IS–150—Comparative Analysis of Processes of Agricultural Development and Modernization).

Professor S. N. Eisenstadt, the principal investigator of the grant, for his encouragement and advice.

Our research staff, who collaborated with us in the collection and the analysis of the data: Mr. U. Almagor, Mr. A. Farber, Miss M. Korelaro, Mrs. A. Peri, and Mrs. M. Shapira.

Professor R. Dore, of the London School of Economics, for a careful reading and detailed comments on an early version of the manuscript. Professor Dore's remarks were of particular value to us not only because of his sociological acumen and insight into agrarian problems, but because he provided the fresh and objectively critical point of view of an outsider, free of the many biases and preconceptions characterizing those intimately connected and identified with the subject.

Keren Hayesod, Foundation Fund, for the air photographs.

The Department of Geography of the Hebrew University for preparing the various maps.

And last but not least, Mrs. S. Weintraub, the wife of one of the authors, who took great pains in editing the various pieces and translated parts from the original Hebrew.

D. W.
M. L.
Y. A.

February 1968

Contents

Illustrations

PLATES

MAPS

CHARTS

Tables

Introduction

This study examines the major forms of Jewish agricultural settlement that emerged in Palestine under Turkish and British rule, and analyzes the main social factors related to their creation and development.

Our examination of patterns and processes of colonization focuses on three basic types of rural community: the *moshava*, the colony of privately owned and individually operated farms; the *kibbutz*, the collective settlement (more precisely, one of its several varieties); and the *moshav*, the smallholders' cooperative village. The study draws primarily on contemporary documents, but it has not been our purpose to trace and describe the historical process as a whole. The point of departure was rather analytical-comparative: we have concentrated mainly on a set of specific social problems that we believed to be of general significance.

The main themes can be summarized as follows:

1. *The character and the significance of the original colonizing impetuses or drives that produced the different settlement patterns.* The way in which the initial push to create a new rural form of life crystallized is approached in terms of the social properties of the founding fathers. That is, we examine: the composition, scope, cohesion, and internal organization of this group and its capabilities for sustained action; its position in the social structure and its command over different types of resources; and the intensity and nature of the group's commitment

to its goal of a new way of life, its image of this goal, the values this image articulated, the problems and interests that brought it into being, and in particular whether and in what way the settlers' commitment constituted an ideological transformation. Finally, we study the concrete policies and institutional arrangements in which the idea was actually formulated.

2. *The over-all setting and conditions, or the societal potential, under which the different settlement patterns were established and developed.* The study investigates the extent to which the social factors with which these settlements interacted and upon which they depended were geared to sustain, modify, or impede—initially and over the course of time—rural development as a whole and the several distinct colonizing drives in particular. These factors are viewed in a twofold way: the limitations of the available resources in general and of the existence of the requisite infrastructure—political, administrative, integrative, educational, and other; in a more directly relevant sense, the degree to which the different rural patterns fitted into the existing institutional arrangements and injunctions, constituted a response to basic social problems and needs, and suited the goals, interests, and policies of the strategic and resource-rich elements, groups, and élites.

3. *Patterns of settlement and their growth and performance.* The issue here is the performance and achievement of the various colonizing systems. Reference is made to the basic patterns institutionalized in the course of the interplay between "initial impetus" and "social potential" and to variations produced over time. Here the framework of analysis is not, however, the process of change itself (which is taken up later), but the crosscut of the main structural configurations that emerged among and within the different settlement types. We do not presume to measure precisely the relative success or advantage of these structures through concrete indices of economic efficiency and profitability. Such a procedure would be dubious methodology [1] and difficult because of the paucity of statistical evidence. The intention is rather to outline the basic institutional and functional

imperatives—agrarian and social—of the three settlement patterns and to evaluate these imperatives, *grosso modo*, in terms of their potential and limitations for development and growth under variable conditions. Of particular interest is the extent to which each of the different forms possessed a sufficient range of institutional alternatives to solve the problems created by their fundamental needs. The problem posed is thus twofold: the differential viability of *specific* configurations in a context of development, and the *over-all* range of structural and attitudinal variability attendant upon and consistent with this process.

4. *The internal capacity of the different systems to change and to develop in a sustained way.* Here emphasis is on how flexible the various settlements proved in admitting and absorbing pressures of novel circumstances and in initiating and promoting new goals, activities, and aspirations. Thus the forms are assayed for the openness of their value systems in admitting and originating new ideas, for the capacity of their basic institutional patterns to recrystallize continuously, and for the ability of their social systems to produce, encourage, and mobilize innovating élites and entrepreneurs in the economic, social, political, and symbolic spheres. The assumptions underlying this emphasis are that growth and development are by no means a self-maintaining process; the already developed and modernized communities and frameworks are also inevitably faced—at least potentially—with the reality and necessity of change.

The focus thus is on the identification of the critical periods when the challenge arose of changed circumstances, whether internal or external, and on the way such challenge was met. However, a vital distinction is made in this connection between rampant or revolutionary change, in which the fundamental pattern and identification with rural life are broken and discarded, and innovation carried out in reference to the basic commitment by way of on-going secondary institutionalization. This seems to be a central problem in a context of rapid and advanced urbanization, no less salient than that of rural stagnation. Particular attention has been consequently paid to the significance of different social and cultural elements and configurations in this respect.

Of course, the quality of flexibility derives in some measure from the breath infused into the experiment at its creation. The founding fathers, like physical parents, provided both the heredity and much of the environment of their offspring. This dimension, however, is not only analytically distinct, but also inevitably assumes autonomous features in the process of growth, maturation, and social change itself.

5. *Participation in central institutions of the society*. This factor is the obverse side of the coin just discussed. It refers to the political and organizational "genius" of the various groups—their ability to outgrow local, regional, and other parochial frameworks and interests, and to promote and combine into national settlers' organizations. The crucial variable here is the extent to which such organizations are able to exert influence upon central institutions and strategic groups of the society (and perhaps outside its territorial limits as well) so as to participate actively in the determination of the goals of the society and in the allocation of its resources. This struggle for influence and power increases in importance, of course, in direct relation to the exigencies confronting the development of various rural patterns, whether internally or in respect to the over-all conditions outlined above. The checkered colonization of Palestine was characterized, in consequence, by an intense struggle for recognition and a fierce competition for resources.

As is clear from our discussion of the initial colonizing drive, the organizational and political genius was also basically influenced by the social characteristics of the founding fathers. This factor is, as it were, a dependent variable to the independent one of their predispositions—the extent to which they constituted a potential nucleus for a larger framework and the degree to which they visualized themselves as active participants and contenders on the general social scene. Over time, however, the settlement movement or organization assumed, as will be seen, life and properties of its own, and the authoritarian relationship between parent and child was reversed. It is primarily in this sense that this theme is given an autonomous and a central treatment, distinct from the historical origins.

These, then, are the main themes of the book. Their concrete dimensions will, of course, become clear as we go on. We have tried to formulate our parochial historical problems in more general terms; to what extent the findings can be validly interpreted in such terms must, of course, be judged by testimony of the data.

The organization and presentation fall into four parts. First, a general survey of Jewish colonization in Palestine is sketched to familiarize the reader with its many specific features and institutions and with the conditions under which it took place.

Second, examples of a colony, a collective, and a cooperative smallholders' village are examined. Each of them is the first of its kind (with the reservation that the kibbutz is a novel variation). Each example thus highlights the significance of the forces that initially fashion a new social experiment, and each community is also broadly characteristic of others that have proliferated in its wake. But no single example can be fully representative of a whole social reality composed of many units, no matter how generally similar and homogeneous they are. The data presented, while prototypal in the true sense of the word and capable of being generalized, are essentially individual case studies.

An examination of the settlers' organizations of the three settlement types then follows, tracing their development, structure, activity in relation to centers of legitimation, power, and resources, and their significance for the member units.

Finally, a comparative analysis is attempted and some general conclusions offered in terms of the problems that have been raised.

PART I

SURVEY AND CASE STUDIES

1

History of the Jewish Settlement in Palestine: A Survey

The present chapter presents a survey[1] of the social, economic, and political background of Jewish settlement in Palestine and describes the institutions that provided the framework for this activity. It is hoped that such a survey will give the reader a better grasp of the various subjects discussed in the later chapters. This background is presented in the briefest possible way and includes only data directly relevant to rural colonization. Thus, for instance, when demographic factors are described, only the waves of immigration from Europe—those which significantly contributed to agricultural development—are enlarged upon. In reality, of course, there was a Jewish population in Palestine before the so-called First Aliya,[2] and there were also waves of other immigrants, mainly from Mediterranean countries, that arrived during the period under review.

The Jewish settlement of Palestine began with what is usually called the First Aliyah. It consisted of a stream of emigrants from East European countries (Russia, Poland, Rumania) who came to Palestine in two periods, 1881 to 1884 and 1890 to 1891. Twenty to thirty thousand immigrants entered the country during that time. The inspiration for this immigration was the Hovevei Zion movement, which arose in reaction to the Russian anti-Jewish riots (pogroms) of 1881. Its purpose was to change the life of the Jewish people through settlement in Palestine, the ancient homeland, where they would engage in productive work and become a socially and economically healthy nation.

The occupations of the immigrants varied widely, and they included small capitalists, merchants, and workingmen. These immigrants established a considerable number of *moshavot*, agricultural settlements, on land which they acquired with the financial help of the Hovevei Zion movement. However, two factors —lack of agricultural knowledge and experience and the paucity of the movement's financial resources—soon brought the moshavot to the brink of failure.

Baron de Rothschild [3] was called upon to help and prevented the disintegration of these settlements in 1886–1888 by supplying financial resources and professional agricultural guidance. Through the initiative of a team of specialists that he sent to Palestine, the settlers began to specialize in vineyards.[4] At the same time, cellars were built for a wine industry and the way was paved for marketing the produce abroad.

There were, however, negative social features to the economic rehabilitation of the settlements. The principle of self-reliance was undermined, for the Baron's financial assistance was not based on the settler's productivity but on the number of persons in his family. Thus, the settlers were relieved of responsibility and had no incentive to become proficient in agriculture. Again, the professional and business administration of the vineyards and the wine industry was concentrated in the hands of the Baron's experts, and the settlers had no voice in decisions. The pioneering social and economic values of the settlers were gradually dissipated, and in their place came apathy and bitterness against the policy of absolute domination by the Baron's functionaries.

In 1899, Baron de Rothschild decided to correct, as far as possible, the economic mistakes made in the settlements he supported. He transferred their administration to the Jewish Colonization Association, or J.C.A., founded by Baron Maurice de Hirsch in 1891, which had hitherto concerned itself with the settlement of East European Jews who had emigrated to the Argentine. From 1900 onwards, the Association concentrated its

Palestine activities in two spheres. One was the administration of the settlements founded by Baron de Rothschild in the Judaea area and in Upper Galilee. In order to make the settlements self-supporting, the existing system of grants was changed to one of credits repayable at a future date. The Association even transferred the management of the wine cellars and the responsibility for marketing the wine to the settlers. Secondly, the Association established new settlements, based on field crops, on land acquired in Lower Galilee. Here too the settlers were offered loans and not philanthropic grants.

In one respect, J.C.A. repeated the mistakes made by Baron de Rothschild. Its settlements too were based on one-crop agriculture, the raising of grain. Although the grain was not intended for export and therefore was not influenced by the fluctuation of prices in the international market, it was affected by climatic conditions, and thus the survival of the settlement was endangered in times of drought.

The Jewish settlement in Palestine initiated by the Hovevei Zion movement and later by Baron de Rothschild had no legal status. At that time, and up to 1919, Palestine was a part of the Ottoman Empire,[5] and the Turkish authorities were opposed to Jewish immigration; thus, an atmosphere of tension was created. The Zionist movement, founded in the nineties by Theodor Herzl, proposed to solve the problem of Jews as a persecuted minority dispersed throughout the world by creating a legal status for the settlement of Jews in Palestine. Herzl hoped to obtain the Turkish government's consent to this plan through diplomatic channels and with the aid of the Great Powers, if necessary, by financial means. At the first Zionist Congress, initiated by him and held in Basle in 1897, his political plan was adopted. So were the principles that, in his opinion, should govern Jewish settlement in Palestine if his political plan materialized. Cooperative enterprises were to be encouraged in towns and villages, and settlers were to be assisted by way of credits and not by philanthropic grants. A study of conditions obtaining in Palestine was to be made before settlers were brought to the country, and institutions designed to execute the settlement plan were to be set up. Accordingly, in the period between the first Zionist Congress and the First World War, the Zionist movement proceeded to establish the following institutions, which played a decisive role in administering the Jewish settlement in Palestine:

The Anglo-Palestine Bank was established in 1902. Most of its shares were originally sold through the efforts of the Zionist movement. The bank was intended to serve as the financial instrument for the granting of credits to develop towns and villages. Designed as a public bank, it functioned practically from the beginning purely on the basis of economic considerations and has become, in the course of time, the most important bank in Palestine.

The Keren Kayemeth LeIsrael (*the Jewish National Fund for Israel*) also founded in 1902, was intended to serve as the financial instrument for the acquisition of land in Palestine. Until the early 1920's, its funds were also used to extend credits for settlement projects, but later its functions were redefined and restricted to the acquisition of land and its preparation for settlement. Credits for purchasing equipment for settlements were provided by another fund.[6]

The Palestine Office, established in 1908, was designed to administer and coordinate all settlement projects in Palestine on behalf of the Zionist movement. The necessity for establishing such an institution can be understood in the light of the organizational structure of the Zionist movement. Since 1897—when the first Zionist Congress was convened by Herzl—congresses attended by elected representatives from most Jewish communities in Europe, America, and Palestine were held at regular intervals, as a rule biennially. The delegates represented various views as to how the vision of Jewish national renaissance could be realized. From an organizational point of view, the Zionist movement was, therefore, a kind of federation based on both local and ideological representation. Deliberations and resolutions relating to the Jewish problem in general, and the establishment of the Jewish National Home in Palestine in particular, were carried out by an executive committee elected by the Congress and reflecting the relative strength of the various ideological trends represented within it. As the permanent seat of this body was outside Palestine, it became essential, when practical work was begun by the Zionist movement in Palestine, to establish a body to supervise these activities on the spot. Thus, the Palestine Office came into being.

Until the early 1920's,[7] this body dealt primarily with various economic aspects of settlement: acquiring land, planning the establishment and equipment of agricultural settlements, and providing professional guidance to the settlers. The work of the Palestine Office was subject to the resolutions adopted by the Zionist Congress. In practice, however, by being on the scene and having to react to the pressures exerted formally and informally by various groups of settlers, it became a force in its own right, adapting the directions received from abroad to the needs and attitudes of the settlers. Understandably enough, the Congress and its executive body were not always able to appreciate the motivations and values of the settlers themselves.

It may be noted here that Herzl's political plan to obtain the legal consent of the Turkish government to Jewish settlement in Palestine was unsuccessful. An impasse was thus created in the Zionist movement. Herzl considered the principle of legality as a *sine qua non* and was even willing to accept another territory for Jewish settlement, if it could be done legally. (The British colony of Uganda was proposed for this purpose.) This problem diverted for many years the Zionist movement from planning the establishment of the National Home and left it unprepared to cope with the new wave of immigration to Palestine that began in 1904. Thus, when settlement work supported by the Zionist movement began in Palestine immediately after the First World War, lack of experience, ignorance of climatic and soil conditions, and to a certain extent, the shortage of financial resources lent it a trial-and-error character. On the other hand, this unstable situation made it easier for the new immigrants to impose their own ideological imprint on the social and economic patterns of the settlements formed at that time.

Even when the settlement policy to be followed in the future became clearer, the Zionist movement and its executive body in Palestine were prevented from making long-range economic plans. This handicap was inherent in the structure of the movement, which depended upon voluntary contributions from Jews in the Diaspora to finance its economic and social enterprises. These contributions were irregular and uncertain, and their flow was governed, to no small extent, by international events that affected the ability of world Jewry to support Zionist activities. Thus, no table of economic priorities could be formulated in advance.

Though the proposal to accept Uganda instead of Palestine as the territorial base for Jewish settlement checked practical activities in the Zionist movement until it was finally rejected, it stimulated interest in the Zionist idea among wide strata of the Jewish people and was one of the factors in the emergence of the Second Aliya. This wave of immigration consisted of thirty-five to forty thousand settlers who arrived during the years 1904 to 1914, many of them young middle-class emigrants from Russia.

Many of these young people had been influenced by Russian socialist ideas and had taken an active part in the abortive revolution of 1905. However, the wave of anti-Jewish riots at that time in Russia made them realize that the solution to the Jewish problem did not lie, as they had hoped, in joining the ranks of Russian revolutionaries. They gradually came to believe that salvation lay in combining Zionism with social justice: building a Jewish national home in Palestine and basing it on firm socialist principles.

The value of self-labor was of crucial importance in the ideology of Second Aliya immigrants. They contended that self-labor alone would confer a moral right to the country and its soil, which was at that time—and had been for centuries—settled by non-Jews. They soon encountered obstacles in this respect, however. The work available in the existing Jewish settlement was done by Arab laborers who contented themselves with wages much lower than those asked by Jews. Jewish workers thus found themselves superfluous in the existing settlements.

This state of affairs, combined with another event, gave rise to the idea for a new type of agricultural settlement, the collective, or *kvutza*,[8] which became one of the outstanding features of the Second Aliya. From 1905 to 1907, the Jewish National Fund acquired its first tracts of land, and the Palestine Office began to establish training farms on some of it. The young people who started to work and study on these farms were irked by the very strict supervision of the agricultural experts employed by the Palestine Office, and they gradually developed an ambition to become independent. They accordingly requested land and equipment in order to found a kvutza. Their desire was granted, albeit hesitatingly, and the scheme was regarded as no more than an experiment. But in the course of time the kvutza form, which contained many new social and economic features, took root and became a common type of agricultural settlement.

The founders of the kvutza decided on mixed farming and intensive cultivation, since these require less labor and no hired hands; the principle of self-labor was thus secured. In addition, they promised a more assured economic existence than single-crop farming, which largely depended upon prices in the world market or weather conditions.

The socialist ideal, a central value of these immigrants, was achieved in the social structure of the kvutza. It was based on the principle of absolute equality of its members, both as regards the satisfaction of personal needs and the participation of each member in the management of the kvutza affairs. The farmers also maintained that the collective-kvutza, through its development, would also make possible the maximum development of each member of the group; this, however, was conditional upon the members' capacity for self-discipline, the ability to rise above selfish considerations and lead a strictly ascetic life.

However, the extreme socialist principles embedded in the

kvutza structure, which gave priority to the interests of the collective and assigned a secondary place to the interests of the individual and his immediate family, were not acceptable to all the immigrants of this period. Consequently, at the beginning of the twenties, the first experiments were made to create a new type of agricultural settlement embodying, on the one hand, some of the economic principles that served as a basis for the kvutza, such as mixed farming and intensive cultivation, which promoted the principle of self-labor, and, on the other, social principles that recognized the importance of personal and family individualism. The new type of settlement, the moshav, like the kvutza established on land owned by the Jewish National Fund, placed the main emphasis on the family as the principal producer and consumer. Instead of the extreme equality in supplying individual needs—a basic principle of the kvutza—the moshav primarily stressed cooperation in the marketing of its produce and the purchase of supplies. This principle narrowed the gap of inequality between families but did not abolish it completely.

Another pattern that crystallized during the period of the Second Aliya, and which also became permanent, was the formation of highly mobile groups whose object was to safeguard the land acquired by the Jewish National Fund before permanent agricultural settlements could be established on it. This type of unit was necessary in the existing conditions, for the land laws [9] in force at that time made title legal only if the land was actually occupied by the owner. Title was forfeited unless the owners took actual physical possession within a certain period after its acquisition. The irregular flow of national capital made it usually impossible both to purchase the land and immediately to settle it by providing all the equipment required by the farmer. When these two essential steps were carried out simultaneously, sometimes for security reasons (the fear of squatting by Arab neighbors), the equipment and supplies allocated were usually insufficient and resulted in poor earnings and delay in making the units self-supporting.

Thus, the second wave of immigrants to Palestine, composed

mostly of young people imbued with a strong national and socialist ideology, contributed two new types of agricultural settlements: the kvutza and the moshav. In the early stages of their development, for reasons stated above, these new types gave poor economic returns, but they performed other, more important functions: they safeguarded large areas of land under difficult security and climatic conditions, and they experimented with new approaches to agricultural problems, a thing that private capital, seeking security and profits, did not do.

The national capital collected by the Zionist movement was invested in these agricultural settlements, despite the strong objections in the Diaspora to their social structure (particularly to the kvutza). The Palestine Office realized that this type of settler and settlement best answered the particular conditions of the country and would help to attract private capital in the future.[10]

The end of the First World War brought about changes in the political status of Palestine. As already stated, until that time Palestine was part of the Ottoman Empire, which had only recognized the Jewish community as a religious-cultural group with no political rights. At the end of the war, as a result of an agreement among the victorious powers, Palestine became a mandatory territory under British rule. The terms of the mandate, approved in 1922 by the League of Nations, gave the arrangement a legal sanction, recognized the historic connection of the Jews with Palestine and their right to settle there, and legalized the Balfour Declaration issued by the British government on November 2, 1917, which promised to establish a Jewish national home in Palestine.

This new political situation, however, created tensions and difficulties. The administration of the mandate was subjected to contradictory pressures: the demand of the Jews to protect their rights to immigration and settlement as guaranteed by the mandate, and the opposition of the Arab inhabitants to the implementation of this policy. As a result, the Palestine government, practically from the beginning of its rule, began to restrict both immigration to Palestine and the acquisition of land by Jews. This, coupled with continuous Arab attacks on Jewish life and property,[11] handicapped systematic economic development.

By the terms of the mandate, the Zionist Organization acquired a new function:

> An appropriate Jewish Agency shall be recognized as a public body for the purpose of advising and co-operating with the Government of Palestine in such economic, social and other matters as may affect the establishment of the Jewish national home and the interests of the Jewish population in Palestine, and subject always to the control of the Mandatory Power, to assist and take part in the development of the country. The Zionist Organization, so long as its organization and constitution are in the opinion of the Mandatory Power appropriate, shall be recognized as such an agency. It shall take steps in consultation with His Britannic Majesty's Government to secure the co-operation of all Jews who are willing to assist in the establishment of the Jewish national home.[12]

The Jewish Agency thus superseded the Palestine Office, but its scope of activities was much larger; in addition to economic functions, it became the political representative of the Jewish community in the country vis-à-vis the Palestine government and the Arab political organizations. A considerable part of the Jewish Agency's political activity was devoted to the problem of immigration and to the struggle against the restrictions imposed by the mandatory authorities in this and other spheres. Like other institutions of the Zionist Organization, the Jewish Agency was elected by the Zionist Congress, and its composition reflected the relative strength of the ideological groups represented in the Congress.

In 1921, after restricting the tasks of the Jewish National Fund to the purchase of land and its initial preparation for cultivation, the Zionist Organization established a new fund, the Keren Hayessod (Foundation Fund), to finance the essential equipment and supplies required by the settlers. This fund too depended mainly on contributions from Jews in the Diaspora, but sometimes financed its projects by obtaining loans in various financial markets.

The mandatory administration also brought about changes in the internal organization of the Jewish community in Palestine. These changes were mainly structural: the community was recognized as a religious-cultural group only. However, while under Turkish rule its only institutions were religious ones, subject to a rabbi in Istanbul with a representative in Palestine, it now had the right to choose its own institutions for the administration of its internal affairs.[13]

There were two such institutions, the Elected Assembly (Asefat Hanivcharim) and the National Council (Havaad Haleumi) chosen from among the members of the Assembly and serving as the executive committee responsible for carrying out the Assembly's decisions. The Assembly was democratically elected, and its composition reflected the ideological movements and the many different communal groups comprising the Jewish population in the country. The representation of all the diverse elements was made possible by the proportional system of elections adopted by the first Assembly. This system encouraged the various parapolitical and pluralistic groups to appear as independent parties. Indeed, during each of the four elections to the Assembly that took place in the years 1920–1944, there were between twenty and twenty-five slates. Most of these slates, though, failed to command the hoped for support and demands were made to change to the curia system, in which minimal fixed representation would be guaranteed to organizations representing special groups and interests. These demands were voiced by ethnic associations—especially the Yemenite—but supported also by the Farmers' Association, which did not consider its numerical strength and representation in the Assembly reflected its true economic weight and public importance (see Chapter 5). However, the curia system operated only in the third elections (1931), and proportional representation was later restored.

Mandatory law gave the Assembly the right to collect taxes for the purpose of maintaining the community's educational, social welfare, cultural, and religious institutions.[14] Clearly, however, the functions of the Jewish Agency were of greater importance, and from the point of view of politics and the wielding of power, representation in the Jewish Agency was vital. Nevertheless, competition for a voice in the Assembly was no less keen.

The stream of immigration to Palestine, interrupted at the outbreak of the First World War, was renewed in 1919 because of two events: the promulgation of the Balfour Declaration with the conquest of Palestine by the British Army, which strengthened the hope for the legal establishment of a Jewish national home in Palestine; and the Bolshevik Revolution in Russia, in which many young Jews took an active part, before becoming disappointed by the way in which the revolution developed. Under the impact of these two events, a large group of young people, thirty-four to thirty-seven thousand in number, mainly of Russian origin and penniless, immigrated to Palestine between the years 1919 and 1923, constituting what has become known as the Third Aliya.

The members of the Third Aliya, who had undergone an ideological crisis similar to that of the Second, were also committed to the realization of national strivings in the form of a socialist society. This aliya created, first, the Labor Battalion (Gdud Haavoda) which we shall discuss below. Secondly, it influenced the existing types of agricultural settlements and especially the form of the kvutza. As a result of the war, the flow of national funds into the country was reduced to a minimum, and the absorptive capacity of Palestine was accordingly small. To overcome this obstacle, it was suggested that the framework of the kvutza be enlarged numerically and its economic undertakings be more varied. Hitherto, agriculture had constituted the only economic basis of the kvutza; combining agriculture with industry would enable the existing kvutzot to absorb new members and at the same time to accelerate their economic independence. In this way, the large kvutza came into being; it was primarily intended to meet the needs of that period, but in the course of time, it became a permanent type of settlement.

The fourth wave of immigration to Palestine, numbering from sixty to eighty thousand persons, commenced in 1924 and continued until 1929. The demographic structure of this wave of immigration was different from that of the two previous ones, for fewer of the immigrants were single workingmen. The predominant elements were merchants and artisans, most of whom possessed a small capital and came with their families. The majority came from Poland, where the government had imposed restrictions that adversely affected the Jewish minority.

This aliya was absorbed into two main sectors of the economy: first, new citrus-growing settlements were founded by some of these immigrants; second, this aliya gave an impetus to considerable urban development as a consequence of its investments in buildings and factories (mostly small workshops for the textile, leather, and printing trades). This wave of immigrants laid the basis for industrial development; it also gave a strong impetus to the expansion of existing urban areas. Thus, Tel Aviv had about 16,500 inhabitants at the beginning of the Fourth Aliya in 1924, and 46,000 in 1931. In Haifa, the Jewish population in 1924 was 8,000 but by 1931, it was 16,000, while in Jerusalem, the Jewish population grew from 38,000 in 1924 to 54,000 in 1931.[15] All this in spite of the economic crisis that hit the country in 1929–30 and caused considerable emigration. Tables 1 and 2 summarize this urban and industrial development and underline the role played by the Fourth Aliya. More importantly, they show the balanced character of development in the Jewish community in Palestine, with the growth of the rural sector being economically affected and numerically dominated by processes of urbanization.

As the agricultural settlements operated with insufficient agricultural equipment and supplies and lacked the essential conditions for normal economic development, the Zionist Executive tried, in the middle twenties, to assess their economic position

Table 1. Growth of rural and urban population in the Jewish community of Palestine, 1880–1944

Year	Urban	Rural	Total	Urban percentage
1880	23,000	—	23,000	100
1914	73,000	12,000	85,000	86
1931	134,000	41,000	175,000	77
1939	337,000	138,000	475,000	71
1944	427,000	140,000	567,000	75

Source: Y. Ziman, *The Upbuilding of the Country 1882–1945*, in Hebrew (Jerusalem: Institute of Zionist Education, 1946), p. 40. The criteria of "rural" and "urban" are not given in the source.

and to draw up plans for the allocation of everything required for the consolidation of each settlement type. It was, however, difficult to carry out these plans. The need to continue acquiring lands; to cultivate to some extent at least lands already acquired but not yet settled, in order to establish legal tenure; and to deal with the constant claims of workers' groups, which for a long period had earned their living as hired laborers in the colonies, but wanted permanent settlement, all these created an extraordinarily heavy demand on the two national funds and made long-term planning difficult.

We now come to the Fifth Aliya, which commenced about 1930 and continued until 1939, bringing about 230,000 immigrants in all. At first, most of the immigrants came from Poland,

Table 2. Growth of industry and manufacture in the Jewish community of Palestine, 1922–1943 [16]

Year	No. enterprises	No. workers employed	Investment in £P
1922	1,850	4,800	600,000
1930	2,475	11,000	2,234,000
1937	5,606	30,000	11,637,000
1943	7,700	61,000	22,000,000

Source: Y. Ziman, *op. cit.*, p. 72.

the majority of them young, penniless pioneers; but later, immigrants from Germany escaping from Nazi persecution predominated. Thanks to their good professional training (many of them were men of liberal professions), and to the relatively large amounts of capital that they brought with them—capital in amounts hitherto unknown in Palestine—they had a far-reaching influence on the economy of the country during this period. About three quarters of them turned to commerce and industry. Through their initiative, large industrial enterprises were established in the chemo-pharmaceutical, textile, mechanical engineering, and canned-food fields, enterprises whose scale of production and number of employees far exceeded anything that Palestine had known before.

City areas expanded, and in consequence the demand for agricultural products grew. This accelerated the economic development of the existing settlements. Moreover, the technical knowledge of these new immigrants resulted in the development of new methods of drilling and the discovery of new sources of water. This too promoted agricultural development. About a quarter of the immigrants from Germany turned to agriculture and developed a new type of settlement, the cooperative village, similar to the moshav in its economic structure, but much less strict and uniform in its principles.

In this period, especially from 1936 onwards, monies brought into the country by the national funds played a decisive role in settlement. The advantages of national over private capital, especially the possibility of applying it to objectives that were not necessarily economically profitable, now became clear.

Political considerations [17] made it now urgent to acquire lands on a large scale, especially in areas exposed from the security viewpoint, and to settle on the land without delay. But again the funds available were not sufficient to establish fully equipped settlements. The fifty-five agricultural settlements set up during the period from 1936 to 1939, mainly kibbutzim and moshavim, began to function with only a fraction of the manpower ultimately planned for each chosen location. These nuclei were to create the conditions for the future absorption of the full complement of settlers.

From 1939 onward, immigration to Palestine was virtually forbidden by the mandatory government. During the whole period of the Second World War and after, insofar as immigrants entered Palestine, they did so by circumventing the restrictive regulations. In 1948, with the establishment of the State of Israel, immigration was resumed on a large scale.

The five maps which follow illustrate a little of the process of development sketched above and soon to be unfolded in greater detail. Maps 1 through 4 show the different Jewish settlements in Palestine and Israel (in reference to the main cities and the three villages studied intensively) at various periods of time: 1922, 1931, 1948, and 1956. Map 5 is for general geographical orientation and gives the division of the country into the regions referred to in the text.

We have sketched above the history of Jewish settlement in Palestine from the 1880's up to the Second World War. This development took place in, and was influenced by, an over-all institutional and social framework characterized by the following main features:

The various waves of immigration to the country brought different resources and aspirations. The movement was significant first in that each successive group stamped its imprint upon and often dominated whole phases or areas of economic and social activity. However, while this created a deep and lasting sectoralism, with the community divided along lines of ideology, loyalties, and interests, it also assured a regular, though fluctuating, flow of capital, skilled manpower, and a drive to constant experimentation, entrepreneurship, and innovation in all spheres of life. Thus, private capital accumulated within the framework of the colonies combined with the resources of the Fourth Aliya to establish and consolidate the citrus branch. Another part of the capital brought by this aliya, together with that of the first one, gave impetus to initial industrialization and urbanization. By

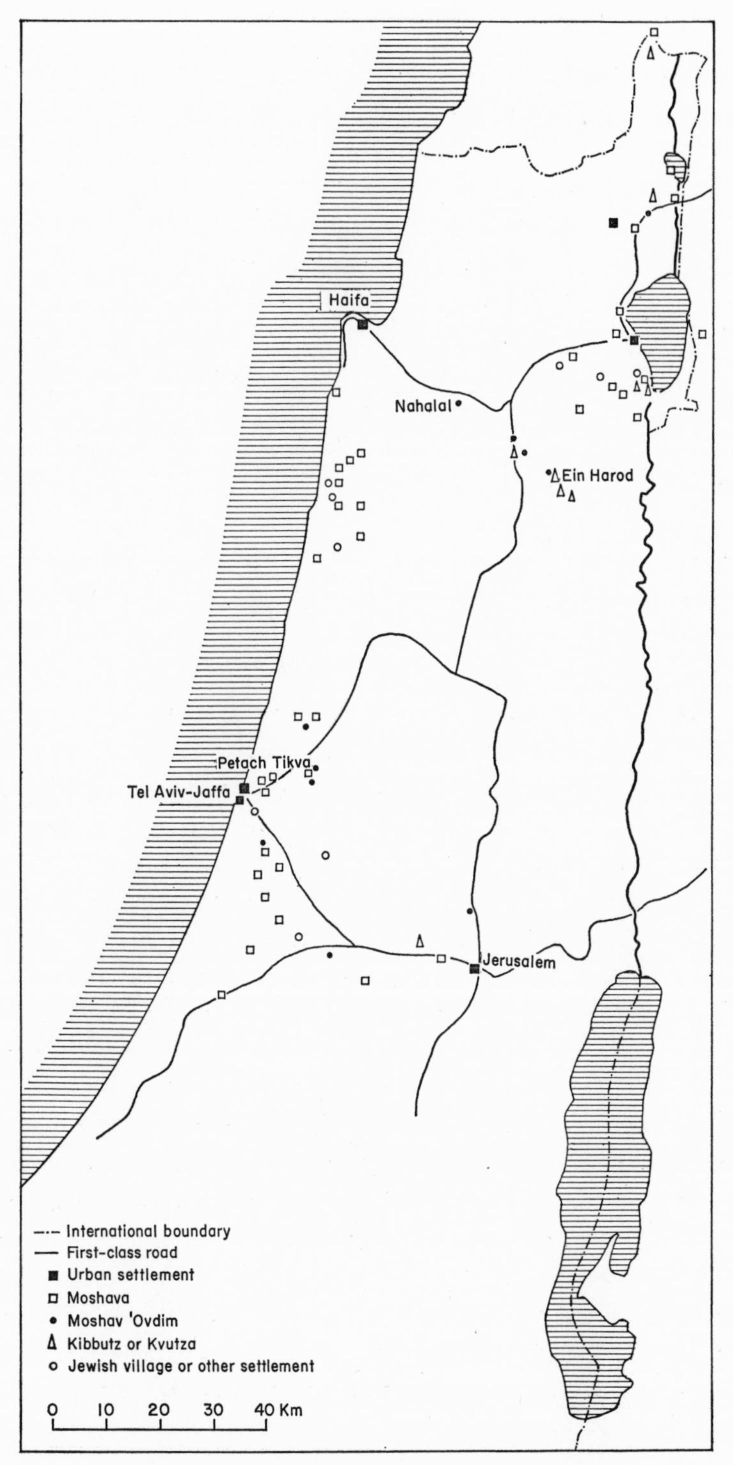

Map 1. Jewish settlements in Palestine, 1922

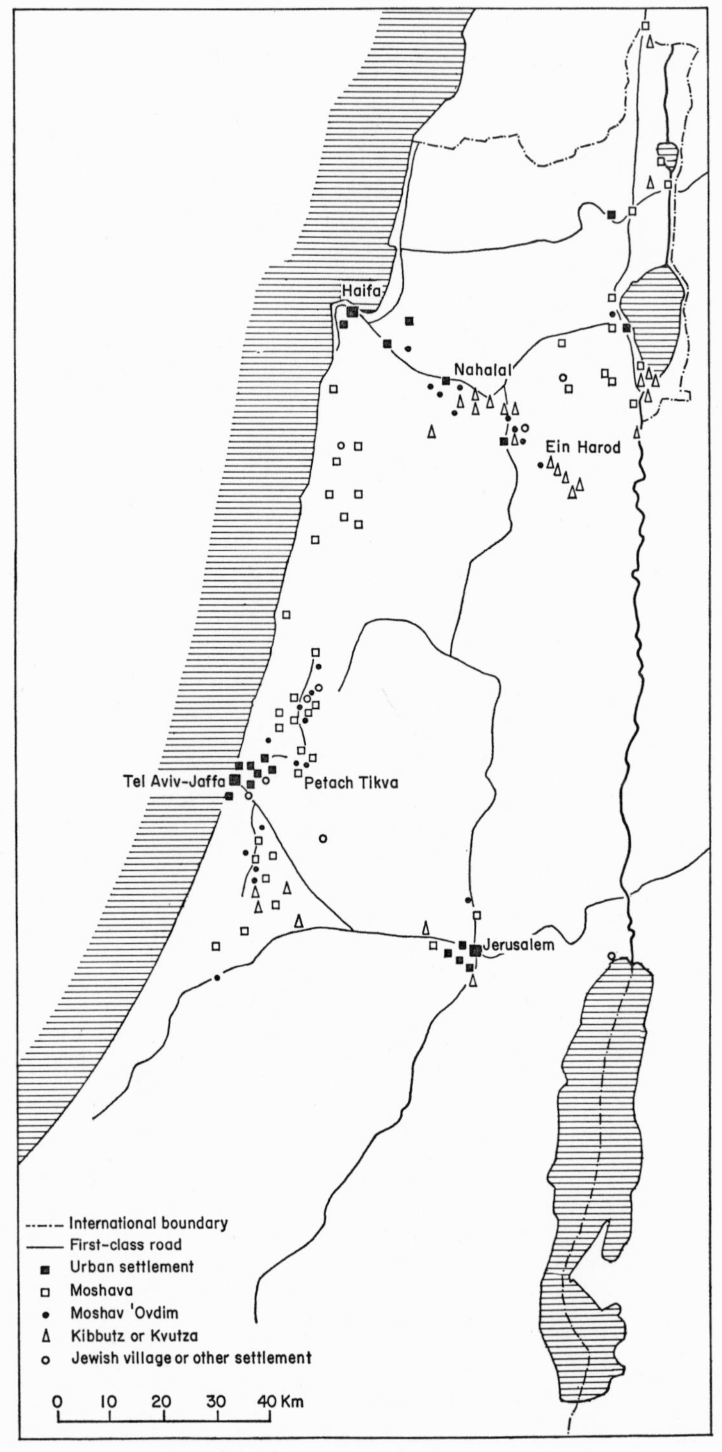

Map 2. Jewish settlements in Palestine, 1931

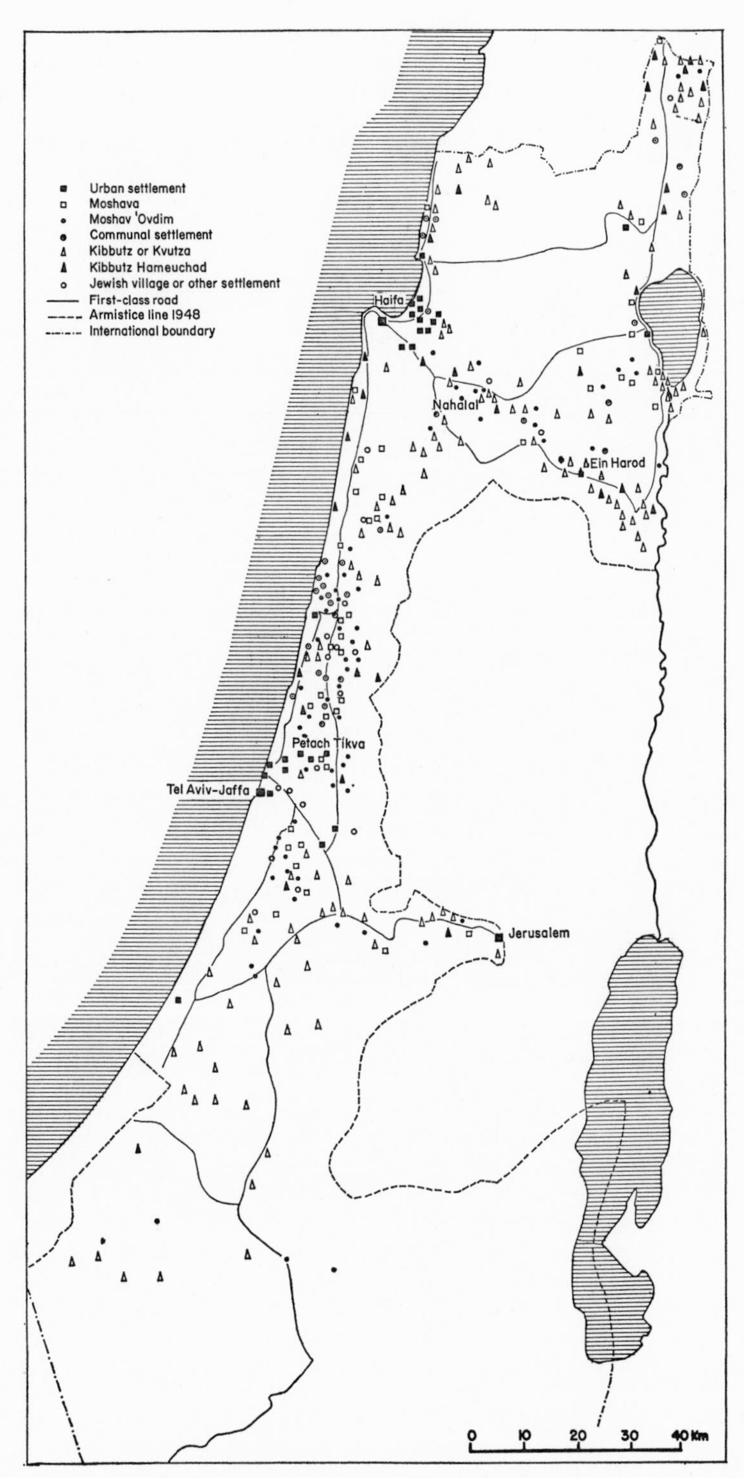

Map 3. Jewish settlements in Israel, 1948

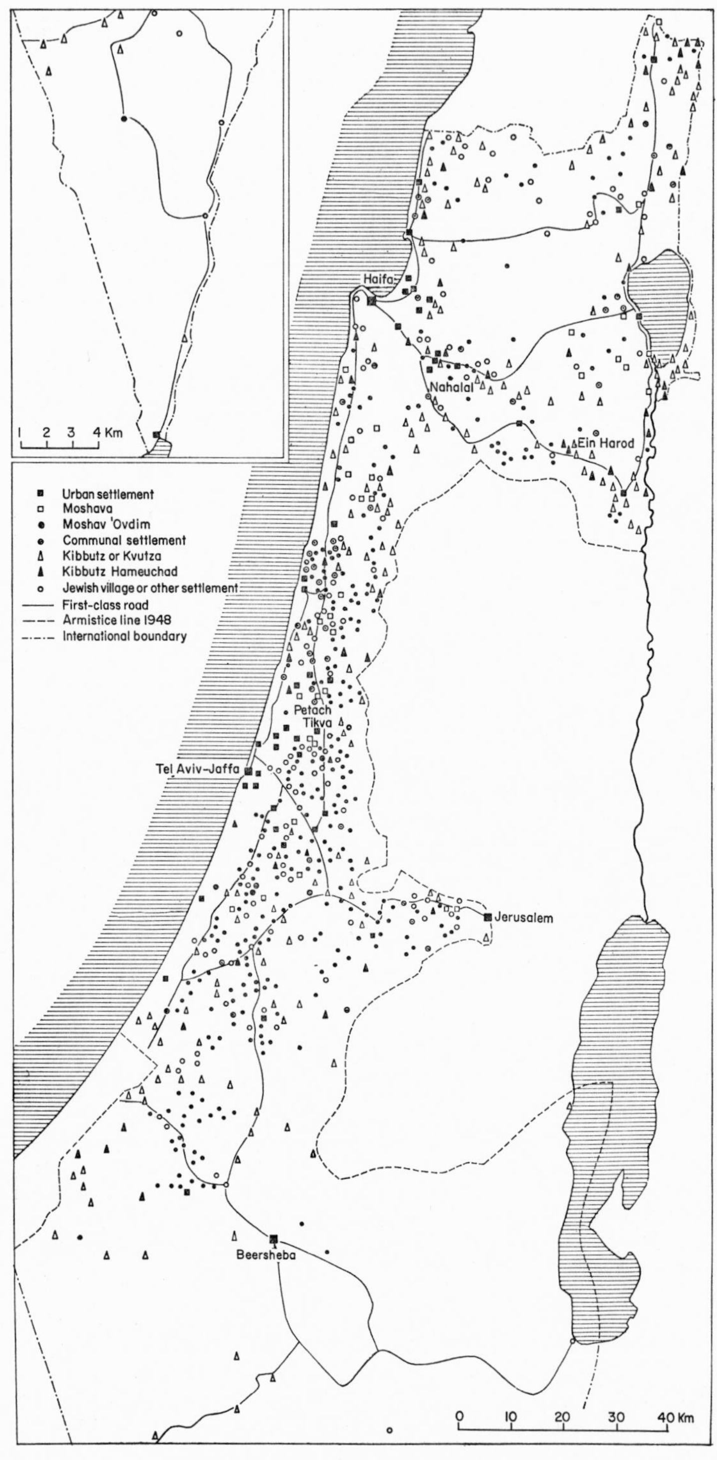

Map 4. Jewish settlements in Israel, 1956

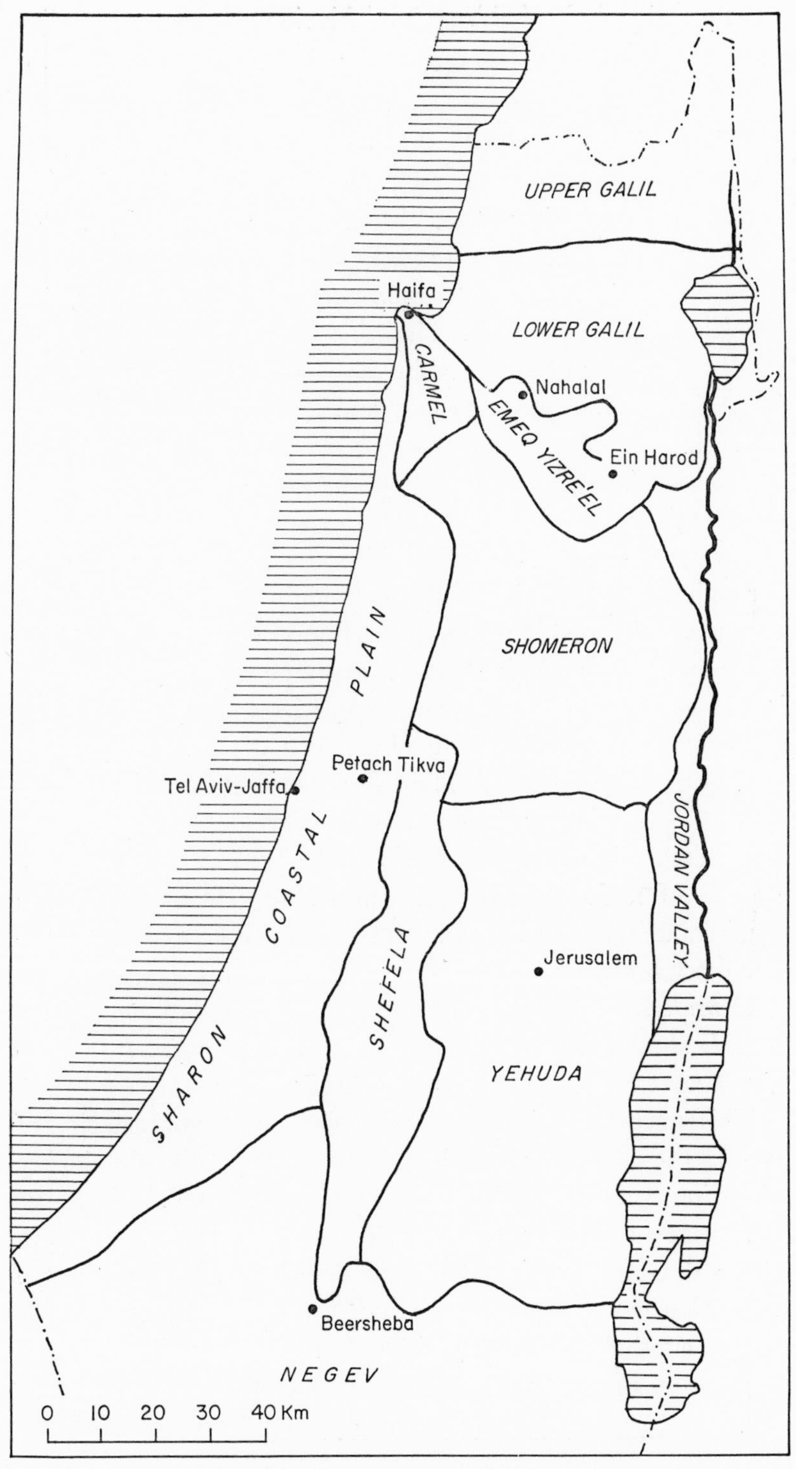

Map 5. Geographical regions of Palestine (Israel)

contrast, the collective and cooperative movements, created and entrenched during the Second and the Third Aliya, exercised decisive social and political influence not only in the rural sector, but also in the Jewish community as a whole.

From the beginning, it was the purpose and the policy of the Jewish national institutions to make the Yishuv[18] *as independent as possible of both the Arab community and the authorities, first Turkish and then British.* This drive towards autonomy extended to such spheres as production and marketing, education, internal political organization and jurisdiction, and social control.

The creation of autonomous and semiautonomous institutions was facilitated by the mandate status conferred upon the country in 1918 under British rule. This status provided a legal basis for action and strengthened the bargaining power of the Jewish community in areas in which the authorities were unsympathetic or opposed to its aspirations. More importantly perhaps, the British mandatory government assumed many of the essential sovereign responsibilities and obligations of external security, administration of order and justice, infrastructure, monetary matters, and others; this not only freed resources for internal development but also gave the Yishuv a chance for a gradual process of nation-building and protected it from the sudden confrontation by too broad a spectrum of problems, a phenomenon often besetting newly independent countries.

Considerable capital was regularly mobilized from sources outside the community through the two national funds, philanthropic contributions, and private investment.

This over-all setting of development does not explain, of course, the differential development and the relative dominance of various rural patterns over the period discussed. It is, as we have seen, an historical fact that the periods of optimum growth of the three forms were seldom contemporaneous. A decline or stagnation in the development of one form often heralded the rise of another. This can be seen from Table 3. In the years 1878–1920, twenty-eight colonies were established in Palestine, between 1921–1945 only fourteen. Prior to 1921 there was not a single moshav in the country; between 1921–45, seventy-nine came into being. Up to 1921 only six kvutzot (kibbutzim) were in existence; by 1945, 104 such settlements were added. Thus, from the quantitative point of view at least, the colonies predominated in the first quarter of the century, while later on the moshavim and kibbutzim became prominent, even though they did not always lead in production and output, a phenomenon discussed below. A comparison between the moshavim and kibbutzim—both of which, as we have said, came into prominence

Table 3. Types of settlements and period of founding

Period	Colonies	Kibbutzim	Moshavim	Total
1878–1920	28	6	—	34
1921–1925	4	10	7	21
1926–1930	4	10	9	23
1931–1935	4	13	33	50
1936–1940	—	40	25	65
1941–1945	2	31	5	38
Total	42	110	79	231

Compiled from A. Bein, *The Return to the Soil* (Jerusalem: Youth and Hechalutz Department of the Zionist Organization, 1952), pp. 422–440.

after 1921—will also show that the periods of development and relative stagnation of each were not always simultaneous. Thus, for instance, in the third decade a roughly similar number of moshavim (sixteen) and kibbutzim (twenty) was built. But in the next five years, thirty-three moshavim and only thirteen kibbutzim were established, while between 1935 and 1940 the situation was reversed, with twenty-five moshavim as against forty kibbutzim being added. This disparity grew larger in the next five years, when thirty-one kibbutzim but only five moshavim were established. Some of these differences could, no doubt, be considered fortuitous. The over-all trend, however, is too marked to be due to chance alone.

The bulk of this book is, of course, devoted to the internal social factors through which this process can in our opinion be explained. These, as we have mentioned in the introduction, are primarily three: the social character, orientation, and resources of the founders, that is, the different immigrant nuclei that established the various patterns; the capacity of the value systems and institutional and social structures of the different settlements to initiate and sustain, over periods of time, innovations and secondary institutionalizations; and the ability of the various settlers' organizations with which the villages were affiliated to influence national policy and allocation of resources.

However, in the present context, something must be said of the main macrosocial factors that, within the setting described above, were significant for rural development as such. These factors, we assume, acted *differentially* upon the three colonizing forms discussed. While their influence was not determinative, and while each of them was modified and utilized in a specific way by the various settlements and their organizations, they nevertheless constituted significant objective stimulants or barriers. Four such macrosocial variables seem to have been of particular importance:

1. *Ratio between private and public (national) capital.* Since the cooperative and the collective settlements were, much more than the colonies, dependent and capable of drawing upon national capital, their ascendancy was assumed to be related to the prominence of this type of financing.

2. *The proportion of land owned by national institutions in the total amount of land belonging to Jews in different periods.* Here, likewise, the kibbutzim and the moshavim were assumed to depend upon and to flourish in proportion to the amount of land at public disposal.

3. *Security conditions in the country, in respect to the intensity of Arab marauding and disturbances.* Since the kibbutzim constituted, as will be seen, social mobilization systems, they were least vulnerable to bad security conditions and most valuable as a defense against them. A high incidence of organized depredations was thus assumed to be conducive to the intensification of collective settlement.

4. *The state of the employment market in towns and colonies.* As will be shown later, membership of moshavim was recruited primarily from among landless agricultural and other workers in towns and colonies. Increased unemployment was consequently considered a significant factor in the establishment of this particular form.

However, the substantiation of some of these assumptions is empirically more difficult than might be supposed. The relationship between type of available capital and predominance of certain rural types is especially hard to evaluate. First, the data for each separate year are not always available. Second, it is difficult

Table 4. Jewish capital imports (in millions of pounds)

Year	National capital	Private capital	Ratio of private to national capital
1898	0.7	6.0	8.5
1920	2.1	13.4	6.5
1925	4.0	24.6	6.0
1930	4.0	16.0	4.0
1935	4.1	42.4	10.5
1940	8.4	27.6	3.0
1945	32.6	59.2	1.5

Prepared from data published in A. Olitzur, *National Capital and the Development of Palestine 1918–1937*, in Hebrew (Jerusalem: Keren Hayessod, 1939), Chapter 10.

to estimate whether the *ratio* between national and private capital or the *absolute amount* of each is decisive. Possibly both should be taken into account and the relative importance of each evaluated. In any case, the data show that throughout the entire period private capital predominated, although the share of national capital grew constantly (except for the years 1935–36, when the greatest number of capitalists, mainly from Germany, came to Palestine). Up to the twenties, private capital was about 8.5 times larger than national capital (see Table 4), and the absolute amount of national capital was small, only 0.7 million £P (Palestine pounds); in that period mainly colonies were established. By the middle thirties, private capital was only four times larger than national capital, and national capital grew fourfold, while private capital only tripled; in that period the number of moshavim and kibbutzim grew considerably. The middle thirties are an exception; in those years the flow of private capital intensified and was ten times larger than national capital, an all-time high; interestingly enough, in that period a record number of kibbutzim was established. This can be explained, however, by a very sharp deterioration in the security situation.

In the forties, the gap between national and private capital almost disappears. National capital becomes abundant in absolute numbers too, and we witness a great development of the moshav and kibbutz forms.

To sum up, before the First World War private capital as well as some national capital was invested in the establishment and consolidation of colonies. After the war, some of the private capital was used to build new colonies, some was employed in strengthening the existing ones, but the largest part went to urban centers. To a considerable extent, national capital was used for the development of cooperative and collective agriculture; it should not be assumed, however, that this happened automatically; pressure was exercised and struggles undertaken to achieve this end. We shall discuss the matter in detail in the following chapters.

Similar conclusions arise from an examination of the ratio between privately and nationally owned land, that is, land bought and owned by Jewish national institutions (Table 5). Before the First World War, private land was twenty-four times larger than national land. This changed rapidly in the ensuing years; by the end of the second decade the ratio was 1 : 3.5, and in the middle forties 1 : 1.1. The rate of increase in nationally owned land is not always faithfully reflected in the rate of growth of cooperative and collective settlements, but the overall tendency for the two phenomena to be related is quite clear. It is, however, important to note here that kibbutzim and moshavim were founded even when nationally owned land and capital were scarce in relation to the number of prospective settlers. The relative superfluity of candidates in certain periods was mainly due to the activities of the settlement movements in Eastern European countries. In consequence, the prospective settlers organized into so-called labor battalions, which had to wait five to seven years before land was allotted to them. Overcoming the disproportion between the number of prospective settlers and the amount of available land was one of the main struggles of the various kibbutz movements.

Table 5. Land in Jewish hands (in acres)

Year	National land	Private land	Ratio of private land to national land
1915	4,100	105,050	25.6
1927	49,175	176,500	3.5
1936	92,450	255,700	2.6
1945	204,300	240,275	1.1

Compiled from David Gurvitz and Aaron Gertz, *Jewish Agriculture and Agricultural Settlement in Palestine*, in Hebrew (The Department of Statistics, Jewish Agency for Palestine, 1947), p. 45.

Table 6 shows the relationship between the security situation and the number of newly established settlements of each type. The relationship is conspicuous and clear: in periods of *relative* peace, 92 per cent of the colonies, 62 per cent of the moshavim, and only 32 per cent of the kibbutzim were established.

The factor of the colonies' and the towns' absorptive capacity is difficult to document precisely. There is no doubt that in periods of economic depression, for instance in 1926–9 or at the beginning of the forties, a large reservoir of manpower for settlement in moshavim (and to some extent in kibbutzim) was created. But the relationship here is indirect, because usually a fairly long period of time had to pass from the time when agricultural and other workers lost their employment and the time when they could be settled on their own land.

The configurations of macrosocial factors which worked for the benefit of the various settlement forms up to 1948 can be summarized as follows:

Colonies predominated under normal security conditions, when there was a low ratio of national to total Jewish land, and when there was a relative scarcity of public as against private

Table 6. Establishment of settlements according to security conditions

	Years of good security conditions				Years of bad security conditions				
Settlements	1878–1914	1922–29	1930–35	Total	1920–21	1936–39	1940–45	Total	Total years
Colonies	28	7	4	39 (92%)	1	–	2	3 (8%)	42
Kibbutzim	6	15	14	35 (32%)	4	38	33	75 (68%)	110
Moshavim	–	12	37	49 (62%)	–	21	9	30 (38%)	79

Compiled from *The Hagana Book*, in Hebrew (Tel Aviv: Maarachot, 1954), vol. 1–3.

Note: Security conditions are defined on the basis of evaluations published in the above-mentioned book, which was the official organ of the Jewish Defense Organization (Hagana). Years of bad security conditions were marked by continuous, organized, and large-scale Arab attacks on the Jewish population (1920–21 and 1936–39). During the period 1940–45 (World War II), internal security was physically good, but there was nevertheless considerable tension because of Rommel's threat in North Africa and the publication (in 1939) of the White Paper, which severely restricted the purchase of land by Jews. It was considered necessary to establish villages in strategic areas, such as the Galilee and the Negev (south), so as to anticipate possible partition of the country after the end of the war.

capital. Factors of unemployment were not highly relevant for this type of settlement.

Collectives were favored by severe security conditions, availability of national land, and by a flow of public capital. However, the kibbutz could function viably on less investment than that required for the minimal capitalization of separate family farms; it therefore had an advantage over the moshav when public capital was scarce. The same was true also of land: the greater flexibility of cultivation in a collective unit of production meant a better possibility of operating where small parcels of land of poor quality were available for settlement. Finally, the employment capacity of towns and colonies was here of secondary importance.

Smallholders' cooperative villages emerged chiefly in normal security conditions, when there was a high ratio of national to total Jewish land, when there was a high ratio of public to total Jewish capital, and when unemployment in colonies and towns was high.

It thus seems that the relationship between macrosocial factors and periods of growth and recession of different types of settlement is clear, even if not fully substantiated empirically. As has been mentioned, however, this relationship should by no means be considered natural or inescapable. It was but a limiting factor, in response to which each village and parent organization exercised its particular talent. It is to the discussion of these talents that we must now turn.

2

A Colony: Petach Tikva

The First Period: 1878–1881

The origins of Petach Tikva go back to a society founded in Jerusalem in 1875 "for the cultivation of soil and redemption of the land." This society was organized by a group of men from the Old City of Jerusalem, some of them born in it, others of Hungarian origin. Practically all of them were town-bred and had no farming experience; they were not youngsters, but heads of large families. Their roots in the traditional and ultraorthodox Jewish community of Old Jerusalem determined their self-image and influenced the definition of their aims, their relationships with other groups, and the patterns of their organization.

The aims of the group were stated in a proclamation issued in 1875: "To settle the land of Israel by tilling the earth and getting our daily bread from it." Further on we find the basis for this idea: "The Jewish people cannot exist without a country. Every Jew should make it his heart's desire to attach himself to the Holy Land, whether in spirit or in deed. . . . The Jew can only feel that he is the descendant of his holy forefathers if he loves his land." In a letter to Rabbi Auerbach in Jerusalem we find the following: "We have decided to buy, with God's help, a plot of land which we shall cultivate and on which we shall live according to the precepts of the Bible. And thus do we intend to put a new life into our families. Laboring on this land is good for body and soul alike." [1] The idea of redemption was an old one, but the special form it took here and the channel for its realization were new: to redeem, at the same time, the Jew from the evils of life in the Diaspora or of life on charity in the Old City of Jerusalem, and to redeem the land from its barrenness. This double redemption would be attained by Jews tilling the soil of Israel. This aim had two aspects: settlement of the land of Israel conceived as a religious obligation, and farming as a way of life.

Both these aspects grew out of the pioneer's religious view of life: farming and settlement of the land were conceived as a holy religious duty; by being a farmer, the Jew would become a better human being, because his daily activities would be a fulfilment of the commandment of settling the Holy Land. At the same time it was a real innovation. According to traditional thought, redemption would be attained with the appearance of the Messiah; the Jewish people had to possess their souls in patience and wait. Here, on the other hand, the idea was to *do* something to bring redemption nearer. This new conception had its roots in a reform movement among the Jews of Jerusalem, a movement which aimed not only at a scrupulous adherence to religious duties but also at changing those traits of the Jerusalem Jews which were the result of living on charity, of "eating the bread of idleness."

The first step was to leave the Old City and build new, modern neighborhoods outside its walls. The next was to change the source of livelihood, to reject the bread of charity and live by work, to prove that Jews in general, and Jerusalem Jews in particular, could farm. Scrupulous fulfilment of religious duties and working the land thus formed the basis of the self-image. Tilling the Holy Land involved the discharge of religious duties connected with agriculture; thus emerged the image of the religious Jew, reformed in faith and deeds alike. This was stated in the society's rules, which later became the rules of Petach Tikva: "First of all is the fear of God. Let it be known that our society will live by the precepts of the Bible as interpreted by our teachers and sages. Whoever defies the word of the Bible will be liable to expulsion from our society." [2]

The rules provided for a seven-member council that was empowered to decide on questions affecting the members, to buy the land and allocate the plots, and to deal with income and expenditure. The council was elected by secret ballot, with each settler commanding a number of votes equal to the number of plots he had bought.[3] The number of plots held by each settler varied from one up to ten or more, in proportion to his investment, but no one had less than twenty acres. After the land had been bought by the society, each settler became the owner of his proportionate share, but he could not transfer any of it without the council's consent. This limitation was designed to prevent land speculation, in the eyes of the members, a profanation of the Holy Land.

Two rules dealt with one of the most important principles, the obligation to live with one's family in the settlement, and the obligation to work: "Each one of us entering the society must himself dwell there [in the colony] with his family, on the piece of land which he holds"; and "Work shall be done by the member himself as long as he is capable of doing it, and no laborer may be hired instead." If, however, the owner is incapacitated, he should "settle somebody else instead who will discharge all the duties as specified in the rules." Another paragraph states: "If a member does not know how to till the land, another one, more proficient than he, will help and teach him; should this cause him to neglect his own land, he may hire a laborer in his stead, so that he should not be the loser and the first one learn too." (This is taken from the statutes of another group which came to Petach Tikva somewhat later).

The settlement was to be carefully planned by an architect, the houses built close to each other:

> After the division of the fields, an architect will be entrusted with the task of drawing up a plan of the village, including marketplaces, streets, and yards—in proportions agreed upon by the majority of the members. Every house will be built in accordance with the architect's plans, and no changes will be made in the fronts of the houses nor in the backyards except in special circumstances. The members

who will build the first houses will put them up close together and not widely dispersed.[4]

Another source relates how the settlement was modeled upon those of Cherkessian villages in Palestine:

The plots were divided up on the plan of a Cherkessian village: the houses surrounded the settlement on all four sides. In the free space thus created in the middle, cowsheds and barns were set up. Only a few gates led to the inside of the village so that by closing them the village could be more easily guarded against attacks.[5]

The second matter to which the settlers turned their attention was education: "The second duty is to provide for the children of school age so that they should not run wild and get into mischief. Therefore, a teacher will be appointed to teach them and look after them." [6]

Mutual help was a basic principle of the new society, whose members were well aware of the difficulty of the task they had set themselves. The rules defined and elaborated this principle in concrete terms. For instance, every man was obliged to set aside a sum each year for the purpose of helping the sick.[7] Specific regulations concerning hospitality were also included. The settlers were obliged to set aside and furnish a house for the use of guests, and there was an obligation to extend hospitality to poor guests. Another rule provided for the appointment of a special official to alleviate sickness or poverty. But although mutual help was institutionalized and represented as a duty, it was still on a philanthropic basis and depended on each individual's good will. There was no organization functioning as a society for mutual insurance nor any universal criteria defining the right to seek and obtain help.

An area of 850 acres was purchased in an uncultivated district not far from the port of Jaffa, in the Sharon Valley. It was inhabited by an Arab tribe who sporadically cultivated a small part of the land and grazed cattle. Though not very far from Jaffa, this area, owing to a lack of good roads, remained wild and desolate; the Turkish authorities were incapable of imposing law and order. Foodstuffs and medical help were hard to obtain; in wintertime, when the rains turned the dry gullies into rushing streams that flooded the surrounding countryside, the whole district was practically cut off from Jaffa.

The land itself, though neglected for generations, was fertile. The Yarkon River, which bordered the area, supplied abundant water, and the Jewish settlers planned to grow crops that needed large amounts. However, in this case the blessing turned into a curse: swamps extending along the riverbanks harbored malaria-carrying mosquitoes, and soon the settlers fell victim to the sickness.

It cannot be said that the founders were blind to all the obstacles, but their enthusiasm to embark upon the realization of the ideal of national rebirth drove them onwards. One of them writes in his memoirs:

> On the fifteenth of Ab [a summer month in the Jewish lunar calendar] we set out again to determine the exact spot on which the houses would be erected; we had decided on a hill. When we sighted it, I spurred my horse and reaching the top, slid down, knelt on the ground, and kissed it and watered it with my tears. My companions did likewise, and then we gave a prayer of thanks to the Almighty who let us live to see that day. I felt I was being reborn.[8]

In 1878 the first Jewish agricultural settlement in Israel was founded. The name given to it, Petach Tikva (the Door of Hope), expressed the striving to turn the desolate valley into a source of hope for the redemption of the whole barren land. It was taken from the Book of Hosea (2:15): "And I will give her her vineyards from thence, and the valley of Achor for a door of hope: and she shall make answer there as in the days of her youth."

At first, only the men came to settle on the land. "They erected a wooden hut on a high hill and began to dig a well. Within two months water was reached and then the sowing began. But owing to lack of housing facilities, they could not bring their families to live with them."[9] All the settlers lived together in one house that they built and where the grain was stored; on Fridays they travelled to Jaffa to spend the Sabbath with their families.

In the first years, land was tilled collectively; work animals and agricultural tools were acquired on a collective basis, and the settlers also lived in common buildings. But detailed accounts were kept concerning each settler's investment. In this initial stage various field crops (wheat, oats, sesame, corn, watermelons, etc.) for the use of men and animals were grown. The choice of agricultural branches was largely governed by value factors: "To bring forth bread from the land."

The collectivism of this first period was not the result of collectivist ideals; it was sheer necessity. The difficult conditions precluded individual undertakings; only a communal group could succeed—if at all—in establishing a viable settlement in the heart of this wasteland.

In this first period, all pleas for help addressed to Jews in the Diaspora were couched in the traditional terms of philanthropy. Emissaries sent from Petach Tikva were given letters of recommendation asking Jews abroad to help the settlers to overcome the manifold difficulties that beset their new settlement. A new note, however, crept in: the Jew in the Diaspora who helped the tillers of the soil in Petach Tikva would share in the blessings of tilling the Holy Land. However, this was but another form of the old idea held by the Jews of Jerusalem that anyone who supported them shared in the blessings of living in the Holy Land. Later, only their own capital was used, and the determination to reject charity in any form made self-financing of prime importance.

The methods of land cultivation as well as of house construction were modeled on the primitive ones employed by the Arabs, although an effort was made to improve them. But this effort was not at first successful:

When it was decided to give up the primitive wooden plow in favor of a better one, David Ringer imported plows from Hungary. But the local oxen had not enough strength to draw them. Accordingly, oxen from Damascus, bigger and stronger than the local ones, were imported, and plowing could proceed apace. The crops were good and the neighboring Arabs were full of praise. . . . However, the Syrian oxen could not stand the heat of the Sharon Valley and died off one after the other.[10]

When the first crops had been gathered, a tithe was sent to Jerusalem with a camel caravan. This was an attempt to revive an ancient custom from the days before the Jews had been expelled from their homeland. Indeed, the Jerusalemites were greatly moved by the sight, and many of them, convinced that the Petach Tikva settlers had taken the right path, decided to join them. A year later the second Jerusalem group came to Petach Tikva.

The new settlers wanted to live near the river, and not on the hill. The older settlers in vain tried to dissuade them. . . . And so they built three stone houses near the river as well as six clay and wooden huts. They planted vegetables, watering them with the river waters, and they drank this water too. Within a few months the area was covered in lush green vegetation, a pleasure to behold; there were also plenty of fish in the stream. But when summer came malaria struck, carrying off so many victims that every single house had somebody to mourn. But worse was to come: with the advent of the rainy season, the Yarkon River flooded the farms and swept away the clay huts. . . . Marauding Arabs took care of whatever was left. . . . Next year malaria spread to the hill settlement as well and took its toll. Finally, persuaded by relatives in Jerusalem, one after the other the remaining settlers went back to the city.[11]

The abandoned site lapsed back into wilderness.

The Second Period: 1882–1887

One year after the abandonment of the site, the original group of settlers came back and settled again. A number of houses were built on a hill named Yahud, at some distance from the malaria-bearing Yarkon swamps, and also at a distance from the fields to be cultivated. Again the settlers began to sow and to plant food for themselves and their cattle, but conditions were no better than they had been in the first period.

The life of the Yahud settlers was very hard. The way from the village to the fields was long and passed through Arab lands; the Arabs often put up roadblocks to prevent passage. The women, town born and bred, suffered most. The baking of bread was a heartbreaking job and a whole day was needed to bake enough for one family. There were no ovens and no dry wood for making fire, so a primitive oven modeled on the Arab one was set up and cattle dung or thorny bushes used for fuel. The flour, bought from Arabs, was of inferior quality. Water for the vegetable garden had to be carried in jars; it was water left over in the gullies and in the swamps. The staple food was boiled beans and corn. The workers in the fields drank the swamp water after sieving it through their handkerchiefs to get rid of insects, frogs' eggs, and leeches. When going to Jaffa, the settlers rode on asses, but during the rainy season the gullies were full of water; they had to dismount, undress, and wade naked to the other shore. In the village itself there was no doctor, no nurse, no medicines. . . . The houses, built in Arab style, leaked. And yet no one left and the settlers clung to the site with their fingertips.[12]

The school was one of the first institutions to be built, in the year 1884. It started with fifteen pupils and two teachers. At first, only traditional Jewish studies were pursued there; later on arithmetic, geography of the area, and handicrafts were added to the curriculum.

The group basis of the settlers was now different, wider. Two groups were now dominant: the men from Jerusalem and the men from Bialystok, a town in Poland; these two groups amalgamated in the course of time, and the first council, elected in 1884, included people from both groups. The change in the group basis by the cooptation of people from abroad was not accidental. The first settlers attributed their failure partially to the rejection of outside resources. Experience showed that their own resources were not sufficient.

A new element in the relationships between Palestinian Jews and the Jews of the Diaspora came to the fore in those crucial years (1881–1882). The pogroms in Russia brought out the urgency of the Jewish problem and stressed the need for its speedy solution. Among Eastern European Jews, especially in Russia, the Hovevei Zion movement emerged, which saw settlement in Palestine as the answer. An emissary from Petach Tikva established contact with the Hovevei Zion branch in Bialystok. As a result, a group of Bialystok Jews came to Petach Tikva in the first year of its reestablishment. It was in Petach Tikva that the meeting occurred between men of the reform movement originating in the Old Yishuv in Jerusalem, as a reaction to its way of life, and the men of the reform movement originating among Diaspora Jews. The return to a life of manual work in Palestine was the basis of the union between the two movements. A reciprocal relationship emerged instead of the traditional philanthropy. The enlargement of the group basis, especially by the addition of the Bialystok group, strengthened the ideological bond on a concrete, personal level. The Bialystok branch of the Hovevei Zion movement was a loyal supporter of the Petach Tikva settlers. But the personal interest of the Bialystok community in families that settled in Petach Tikva reintroduced the philanthropic pattern of support through the back door, in the guise of family help. The two patterns, the traditional-philanthropic and the mutual-modern, were here intermingled.

Financial demands were defined differently: the Diaspora Jews were asked to finance specific plans, as detailed in the memorandum sent by the Petach Tikva farmers to a congress of the Hovevei Zion movement; they were no longer asked for donations for unspecified matters. Even more important, the Petach Tikva farmers asked for loans; outright donations were to be henceforth used to finance public projects only.[13] True, the principle of financing through loans was not always adhered to, but its very appearance is important. Following upon this aid, conditions changed: housing improved, wells were dug, and various communal services were set up, such as rudimentary medical help, more or less regular transportation, and schooling.

The new relationships between Palestinian and Diaspora Jews provided a legitimate basis for the latter to control in some degree the use to which their money was put. The plans had to be detailed; the congress also decided to send emissaries to Palestine to see for themselves how their money was spent and what were the settlers' needs.

Up to this period, the institutions of the Old Yishuv had a monopoly on contact with groups abroad and exclusive control over the money they sent. The contacts established by Petach Tikva farmers with groups abroad led to an acute conflict with the Jerusalem community. In 1885 a proclamation was issued against the colonies and against the idea of settlement in general. The authors openly admitted that the establishment of the colonies led to a reduction in the amount of charity money and thus restricted their means of livelihood.[14] They argued that the colonies wasted public money and were a source of heresy. They attacked the very idea of colonization and called it a delusion and a lie. A second proclamation came later from the same source, this more moderate in expression and endeavoring to bring about a reconciliation. The separation between the Old Yishuv and Petach Tikva, however, was by then a *fait accompli:* the new colony, whose members came from the Old Yishuv and continued its traditional religious way of life, had by then transferred the focus of their ideological and organizational relationships from the Old Yishuv to the national movements that arose abroad. This, however, did not change their religious outlook; the Hovevei Zion and Petach Tikva people alike remained religious conservatives.

The change in the nature of the relationships with groups abroad, and thus also in the patterns of financing, was even more conspicuous in the next stage of the development of the ties with Hovevei Zion: a permanent committee was established, with members including Petach Tikva settlers and Palestinian Jews outside the colony. This committee decided how the Hovevei

Zion money was to be spent. In 1885 the committee unanimously decided to stop support to individuals. "Henceforth Hovevei Zion will limit their activities as concerns Petach Tikva to matters of general import, such as the preparation of planting and other things likely to promote the colony's welfare and encourage its development."[15] This committee was the institutionalized financing agency, with authority and power of decision, and its financial policy was clear cut. The common ideal of Hovevei Zion abroad and the settlers in Palestine created an organizational instrument adapted to its needs: a financing organization that rejected the philanthropic pattern and adopted instead a modern investment policy.

Although not yet a business partnership or a profit-yielding investment, it was already an impersonal contribution based on reciprocity and partnership in a common task, a selection of expenditure items, and control over *de facto* spending. This was the first framework common to the Palestine community and the Diaspora movement: the council members were appointed by Russian Hovevei Zion from among the settlers in supported colonies or people associated with the settlement of Palestine.

However, while this institution successfully defined its aims and principles of action, it was unable to institutionalize its organizational structure. There were no clear methods of selection or rotation, and resignation of one of the central members broke up the whole council. The empty coffers of the Hovevei Zion at this time drastically reduced the help extended by the movement to the settlement of Palestine. Its place was taken by Baron Rothschild.

He had turned down the first pleas for help because his policy was to have full responsibility for the colonies he supported. This entailed the transfer of management functions from the farmers to the Baron's functionaries. The Petach Tikva settlers, however, who were still bound to the Hovevei Zion, only applied for partial support for specific purposes. However, as the resources of Hovevei Zion dwindled almost completely in 1888, the Baron began to support twenty-eight families; another twenty-eight farmer families did not receive support. The help of the Hovevei Zion council, which existed up to 1914, continued to a certain extent, but the decisive factor was the Baron.

Compared with the help extended to the colonies by the Hovevei Zion movement, the administration of the Baron's officials constituted a regression to the personal-philanthropic pattern. Somewhat later, when Rothschild himself realized the economic and social drawbacks of his system of personal help, he transferred ownership to the J.C.A. and through them to the settlers themselves.

As mentioned before, the close contact with external bodies, first with Hovevei Zion and then with Baron Rothschild, made the colony independent of the Jerusalem rabbis; the center of public activities shifted from internal institutions to external ones. First the council that represented Hovevei Zion became the central institution in the life of the colony; later on the Baron's officials began to govern it. Both agencies were, in different degrees, external to the colony, as against the internal institutions (colony council) of the previous phases. However, the Hovevei Zion council at least associated the settlers in its decisions, two of its members being Petach Tikva settlers, while the Baron's officials were autocratic: all authority was in their hands, all benefits were distributed as they saw fit, and the settlers became hired laborers on their own land.

The village council became an institution whose main purpose was to cooperate with the Baron's officials. Up to 1896 the council members were drawn from among the twenty-eight supported settlers exclusively. In addition to financial and social advantages, these supported settlers also enjoyed formal power positions. Relationships with the officials also determined unofficial status in the colony, and the supported settlers looked down on the "rejected" ones. The famous writer Ahad Haam, during his visit, expressed his horror at the fact that the title "colonist" was given to those farmers who were dependent on the Baron and not to those who earned their livelihood by their own efforts.[16]

In the period from 1882 to 1886—from reoccupation of the land up to the Baron's intervention—the agricultural patterns were similar to those of the first period: mixed field crops for own consumption. This type of extensive cultivation could not supply all the settlers' needs. At that time it was thought that an increase in the size of cultivated land—up to 37.5 acres per family—would enable the farmers to live on their farm revenues, without any external support.

Most families had 12.5 to 25 acres each, but there were great variations: the biggest holding had 225 acres and its owner lived in Europe. The problem of distance to work had been solved by new houses built nearer to the fields; but another problem arose: absentee landlords and consequent neglect of many plots, with the cultivated ones becoming increasingly rare.

The Baron's experts, however, decided that economic independence would be attained by changing the type of agriculture from field crops to fruit farming, which yields a higher income per unit of land. Since there was no inner market in Palestine, the produce was intended for Europe. Because of transportation difficulties, the wine branch was chosen as the most suitable. Vineyards require a large initial investment, and this Rothschild was prepared to supply: he financed the planting of the vines and the building of the wine cellars; he also ensured the livelihood of the farmers until the vines should start producing fruit. He improved the social services, provided medical care, and modernized the school.

The conspicuous rise in investments in the Baron's period brought about a doubling of population within ten years (from 54 families numbering 287 persons in 1886 to 130 families numbering 600 persons in 1896). The amount of land remained fixed after 1886 at 3,500 acres; the increased investment per acre, however, increased the intensity of land cultivation.

The Hovevei Zion movement, like the Baron, sought to make the farmers financially independent. An emissary was sent to Palestine to examine the needs of the supported colonies and the way money was being spent. He was also empowered to nominate a council to organize the distribution of money and supervise expenditure. In his letters[17] he summarized the basic shortcomings of the colony: "A bad end can be expected from this support, which weakens the individual's strength and makes him lazy and indolent."[18] The general money subsidy paid out each month to the farmers killed every sign of initiative and encouraged idleness. Assuming that the plots were too small to enable the farmer to live on field crops, the emissary reached the conclusion that the farmers should be helped to plant orchards, not to the exclusion of all other crops, but only as a supplement. It was decided to provide a basis for the livelihood of farmers' families by means of one large investment that would ensure their independence later on. The branch chosen was wine-growing. Both these decisions were similar to the Baron's policy; however, because of its scarcity of resources, the Hovevei Zion could not implement them.

The Baron did, in fact, supply the necessary capital, and a new wine industry was thus created, to which agricultural production became geared. The "rejected" farmers switched over to vineyards, too, by recruiting private resources or with the help of Hovevei Zion. The agricultural change thus spread from the farmers supported by the Baron to the whole community. But the wine cellars belonged to the Baron.

The discriminatory practices between the supported and nonsupported farmers applied not only to basic investments and monthly subsidies, but also to marketing possibilities. In bad years in the wine market the Baron's cellars accepted only the grapes produced by the supported farmers. This constituted a serious blow to the "rejected" ones, because they too relied more and more on the vineyards and less on field crops. Finally, they decided to establish cooperative cellars of their own; this represented one of the important enterprises of the period, both economically and socially—the cellars were one of the first productive cooperatives.[19]

This change in agricultural patterns made the farmers absolutely dependent on the Baron. Field and other crops that satisfied domestic needs were supplanted by a commercial branch that required large investments and brought no profit for a number of years. Even then the Baron was the only buyer in good years and the only source of credit in bad ones.

The farmers' inability to gain a livelihood from their farms could not be solved by enlarging the size of each farm, for the amount of land was limited. The solution was to shift from extensive to intensive agriculture, a solution reached, as we have seen, by all the supporting agencies. The strongest push in this direction was given by Baron Rothschild; the first large-scale experiments were his.

But a favorable response to an external demand for change is contingent upon an absence of value or social-structural factors likely to block such change. The family, which was the production unit, was independent; its economic activities were not dictated by extended family relations or communal institutions. However, the motives that had impelled the founding fathers to leave their native cities and to settle on the desolate swampland were not conducive to commercial agriculture. In the conflict between values and new economic possibilities, the latter won.

Thus at the end of the second period the values of the colony shifted. At the beginning, the tilling of the soil was an aim in itself. This aim was conceived in the religious terms of redemption: the redemption of the Holy Land of its barrenness and the redemption of the Jewish people. Settlement would realize this ideal. The proclamation for the Jews of the Diaspora, the rules drawn up by the settlers, the things said by their emissaries, all are aflame with the vision of changing the history of the Jewish people. Sympathizers saw those first settlers as the carriers of national mission; detractors branded them rebels. But both camps viewed this way of living as an innovation: new men in a country renewing itself, redeeming the land at the same time the land was redeeming them.

But the principle of living by agriculture, applied in practice, forced economic considerations upon the settlers. Economic success—not accumulation of property but simply gaining a livelihood—became the criterion for evaluating the principle itself. The settlers and the idea of settlement both had to pass the test of practice, and it was an economic test: Can Jewish agriculture in Palestine be self-supporting? Hence the ascendancy of economic elements over the value elements in the conception of agriculture as the basis of the national ideal and of self-image. This value transformation did not affect other spheres of life: the traditional religious ideas and way of life continued in Petach Tikva for a long time, and changes in this respect only appeared after long and bitter struggles. This was not a change in emphasis on values in general, relative to economic criteria, but an inner transformation of the idea of agricultural settlement. The first settlers wanted, literally, "to bring forth bread from the land," to eat what they had themselves grown on the land. The development of this idea led to a commercialization of agricultural activity and to the use of hired labor, complete contradictions of the original idea.

The original idea found other adherents, but Petach Tikva carried the banner of "self-improvement," economic consolidation of the farmers with the help of public funds. This new idea was not enunciated in any proclamation or rules of association; but it is most conspicuous when we examine the problems that occupied the Petach Tikva settlers most in those days: change of farm pattern, attitude towards financial support, and problems of labor relations.

Disloyalty to the original idea liberated economic motivation, the driving power towards economic growth. From then onwards economic flexibility characterized the Petach Tikva settlers. The initial incentive to shift to intensive agriculture came from the outside—especially the Baron's activity; but the flexibility of the production units and of the new attitudes in the economic sphere made it possible to respond to this incentive and to go on to a complete development of intensive cultivation on the settlers' own initiative.

In 1897 the Baron stopped his patronage and his place was taken for a short time by the J.C.A. This organization wanted to improve the colony's situation by encouraging some of the settlers to leave; it even offered bonuses to those who were prepared to emigrate. Naturally enough, this generated an atmosphere of helplessness and despair among the inhabitants. But the connection with J.C.A. was broken after a short time. The settlers sought for other supporters for the colony, but could find none. The Zionist movement was at that time in a state of inner instability; it had no clear policy concerning the settlement of Palestine and its resources were meager; the Hovevei Zion coffers were empty. Thus, after many years of constant financial support from the outside, Petach Tikva had now to cope with its financial problems by itself. The crisis was severe indeed: since the subsistence agricultural pattern of the first period had given place to commercial agriculture, the farmers were dependent on credits and on markets abroad. When the Baron withdrew his help, both credits and markets became hard to get. Moreover, the vineyards themselves suffered from natural disasters, and the types of vines used were not suited to the climatic and soil conditions. Thus, the question was whether Petach Tikva would survive this crisis, let alone prosper.

Third Period: 1898–1913

The crisis did not put an end to the colony's life; on the contrary, it gave a new impetus to development. The change that the Baron initiated in the colony's agriculture—the shift from subsistence agriculture to specialized farming—remained. But citrus groves were better adapted to local conditions and yielded greater profits. So, after the Baron's withdrawal, the colony shifted from vineyards to citrus groves. But this time the change was the result of the individual farmers' initiative and was financed by their resources. In 1900 the first citrus-growers' association was established in Petach Tikva; it was designed to promote the initial large investments required for planting the groves (especially the digging of the well) and to organize the marketing of produce. Petach Tikva not only initiated the organization of citrus groves and the establishment of contacts with new markets in Europe; but it also pioneered agricultural improvements in the quality of seedlings and methods of cultivation in orange groves; indeed, Petach Tikva introduced citrus fruits into Jewish agriculture. As a result, an unprecedented prosperity came to the colony. Its area by 1914 reached over six thousand acres; most of this area was under fruit trees, mainly citrus groves. Economic prosperity led to a rise in the standard of living; a good part of the profits, however, was used for new investments.[20] Another channel of development was the establishment of daughter colonies around Petach Tikva. Some of them later became independent, while others remained bound to the mother colony for employment and for administrative purposes.

The exceptional readiness for economic changes is reflected in the development of fruit-growing. The change did not consist merely in a single transition from field crops; there were several changes, from field crops to vineyards, from vineyards to almonds, from almonds to citrus. Time after time, the settlers uprooted fruit-bearing trees and planted different ones, till the most profitable one was hit upon. This may show a lack of good planning, but certainly not a lack of economic flexibility.

Another typical feature of this period is the rise of various voluntary associations. Agricultural associations were designed to cope with the problems of mobilizing capital for investment in the groves; to study methods of cultivation; to provide professional guidance; and to facilitate marketing. A cooperative bank and a great variety of cultural and youth organizations were also established. The strongest driving force was individual economic motivation; yet the settlers were capable of cooperative organization in many spheres. Cooperation, which in other Israeli types of settlement is based on ideological grounds, was due in Petach

Tikva to individual considerations of economic and cultural needs.

Economic growth led to demographic expansion: in the decade between 1900 and 1910 population doubled and reached fifteen hundred persons. The area of cultivated land did not grow in the same proportion. (Exact data for 1910 are lacking.) The reason was additional intensification in agricultural work and the appearance of nonagricultural branches. An observer writes at the time: "Since then [since the stoppage of subsidies] Petach Tikva has had three years of intense activity motivated by a genuine desire to live a life independent of others' opinion. There is at present no colony in Palestine that can equal Petach Tikva in vitality, initiative, and growth in all branches of work." [21]

In this period, growth is based almost exclusively on internal resources.

Economic development leads and influences development in all other spheres; in other words, social and cultural spheres lag behind the economic one, but finally change under the impact of economic developments.

First, economic development led to demographic growth without institutional-structural change in the social fabric. At this stage, the religious way of life—modeled on life in the Old City of Jerusalem or the life of Jews in Lithuanian towns—was jealously guarded. The public institutions of the colony underwent no change for twenty-five years. Active participation in them was in practice limited to the founding fathers; the right to vote was limited to property owners. The very right to live in the colony was subject to special approval. Thus, the public institutions became an impeding factor. This conflict between economic development and sociocultural stagnation had its beginnings in the second period; in the third it began to cause severe social tensions.

The social stagnation was finally broken by the force of economic development, which introduced two new elements into the colony: new settlers and hired Jewish laborers. The arrival of new settlers followed economic success, while the introduction of large groups of laborers was the result of the transition to fruit-growing. This is intensive farming that requires a large number of workdays per unit of land. Therefore, the colony had to absorb a growing number of workers. The occupations, values, and cultural background of the settlers became more diversified. The new farmers were less orthodox religiously and more modern in their social viewpoint and way of life. The workers were different in other ways: they were hired laborers, and their economic interests often clashed with those of the farmer-employers; and their demographic characteristics and value systems were entirely different. After 1890 the number of laborers—who came to the colony individually—steadily grew.

Unlike the first settlers, who organized in order to pioneer an empty place, those who came later found a framework of absorption ready to hand. In the colony there were always some hostels, other services, and a certain amount of employment. Yet the history of the relations between farmers and laborers in the colonies is one of conflict. Many reasons are usually adduced for this: economic considerations, which led the farmers to employ cheaper Arab labor; the secular attitudes and revolutionary socialistic views of the laborers; their tendency to interfere in the colony's management and their attempts to influence the younger generation.

Even the attitude towards national symbols served as a cause of division. One of the severest conflicts broke out over the appearance of the national flag in a procession, an act that the farmers feared would arouse the anger of the Turkish authorities and bring penalties to the settlers.[22] The mainspring of all these conflicts was the value transformation that occurred when the farmers' motivation changed from the idea of national revival to private economic considerations. The farmers became, like independent farmers everywhere, businessmen assessing profits and losses and concerned with their property above any ideology. The laborers, especially those of the Second Aliya, were inspired by the idea of a national mission, and their criticism was expressed in the name of this mission. The farmers felt uncomfortable, and so the conflict sharpened. "If I were an opponent of Zionism and wished to prove to the Zionists the bankruptcy of their ideal, I would need no abstract theories to do so—I would simply point to the greatest colony in Palestine, Petach Tikva." Thus writes one of the laborers' leaders.[23]

The farmers tried to get the laborers to leave the colony by denying them work and sometimes even the right to rent lodgings in the colony. Faced with such a situation, the laborers had only one weapon: organization. The first to organize were those who came with the First Aliya. These workers' associations were very similar to those of the farmers in their initial period of settlement, their aim being to get philanthropic help, mainly from benefactors abroad, and with this aid to purchase plots of land and settle as farmers in the colony.

But things changed completely with the arrival of the second wave of immigration. This wave stamped the workers' organization with the characteristic idea of social rejuvenation, which had been an essential feature of the workers' movement in Palestine and later found its expression in various cooperative undertakings and collective agricultural settlements (kibbutzim). The workers totally rejected the situation prevalent in the colonies and worked to change things radically, not only their own status in the colony (the aim of the First Aliya workers), but the economic and public organization of the colony itself. In fact, the workers of the Second Aliya adopted the original ideal of the Petach Tikva founders—to live by their own labor—but they invested this ideal with a deeper and more comprehensive meaning: the socialist ideology, which reserves to the working class a special historical role. They felt themselves to be pioneers, the chosen few whose mission would change the history of the Jewish people.

A. D. Gordon, the ideological leader of the labor movement, gave a philosophical-spiritual expression to these aspirations: Human perfection was to be attained by a life of toil and simplicity in the bosom of nature. The ideal of labor became the focus of a new national ideology, which said that land can only be gained by means of work.

Genuine moral conquest, which enables men to live and create cultural values, can only be attained by the farmer with his plow, the laborer with his hoe, the smith with his hammer and anvil, the teacher and his school—by all those millions of people who sow in tears and sweat the creative seeds of cultural life. This applies to great nations—and all the more so to us, who cannot and have no right to buy a homeland with money alone.[24]

The spiritual ferment in the workers' circles did not remain confined to the realm of theory, but found its practical organizational expression. Hapoel Hatzair (1906) and Achdut Haavoda (1919), two organizations destined to become the basis of the labor parties in Palestine, were founded in Petach Tikva.

In addition to political organization, other types of organization also emerged: the commune and the cooperative. Their first seeds were present in the communal life led by groups of laborers who united in order to overcome the difficult conditions. The First World War accelerated their organizational and ideological integration. There were two novel aspects here: piecework and distribution of payment among the group members, and fully communal life, with everything shared and shared alike. "It was an important event in the history of the Jewish labor movement in Palestine that when the worker's situation in Petach Tikva was so precarious, the first cooperative groups of hired laborers were established there, and later on the first commune, which included hundreds of workers in the colony."[25]

But at that period the workers were still outsiders in the colony, and all this ideological organizational activity was confined to their inner circles. Thus the homogeneity that had been the basis for integration in preceding stages gradually disappeared. This process was clearly reflected in two severe conflicts that rent the colony throughout this period: the struggle of the young people against the old, and the struggle between laborers and farmers. The young were some of the new settlers and the veterans' grown-up children, who opposed the council, proclaimed their right to take part in the colony's institutions, and advocated a change in the conservative way of life of the older generation. The laborers, independent of the religious supervision of the old settlers, but also excluded from the political life of the colony, struggled for the basic right of living and working there.

The dual conflicts, which divided the colony along two different axes, prevented an absolute break between the camps. The existence of two conflicts prevented a complete identification with any particular group and created partial identifications cutting across each other. Thus, the farmers' sons sided with the laborers in matters concerning the modern way of life while identifying with the economic interests of their parents and their antisocialistic views. The laborers identified to a certain extent with the struggle of the young people, although they saw clearly that it was not their fight, since the latter strove not for a universal vote, but only for an inclusion of all property owners in the voters' list; and they certainly did not concern themselves with problems of employment. Still, the youngsters' fight was seen by the laborers as the first breakthrough in the traditional conservative set-up—as it indeed was.

The various sources from this period reflect both the great vitality and activity of the colony in the economic sphere and the cultural stagnation that resulted from the rejection of the pioneering values that had provided the original impetus. When these had been abandoned, only the traditional religious ones were left to govern the life of the inhabitants.

While the Baron's administration endeavors to improve and to support the "chosen" settlers, there are others who have not been chosen, who have for the past few years improved their situation with very limited resources, and they live and work without supervisors and managers. For the past three years Petach Tikva has been the scene of vigorous activity and of a genuine desire to live without regard to the opinion of others. And now we find in the colony some eighty farmers, owners of vineyards and citrus plantations who live by their own labor. The colonists supported by the Baron have also changed for the better, and only fifteen of them receive help with the vineyard work and also a little help with everyday living expenses. In short, there is now no colony in Palestine with such vitality, initiative in all branches of work as can be found in Petach Tikva. . . . It has, so to say, that strength which is needed in the struggle for existence.[26]

The historian Joseph Klausner, who visited Petach Tikva in 1913, felt that the colony—as well as the whole Jewish community—were faced with two vital tasks: economic development and value revival. In his opinion, Petach Tikva welcomed the first, but not the second:

We crossed the colony and I was astonished to find shops everywhere. I asked how many there were in all, and was told that, in this village having only two thousand inhabitants, there were thirty-six shops!

The good profits from citrus plantations have made this branch the foremost agricultural one in Petach Tikva. The Petach Tikva man dreams about citrus groves day and night. . . . The Petach Tikva farmer has been infected with the "illness" typical of all normal farmers everywhere: land sickness. Whenever he "smells out" a plot of land that can be bought, he becomes simply delirious. No price and no sacrifice are too great to acquire some land that is for sale. He also understands the importance of improving the quality of the soil: some of the grove owners had bought poor land and turned it into good, fertile soil within a short period of time.

. . . I left the synagogue in a depressed mood. Of course, I see no harm in studying the sacred texts; peasants in other lands drink wine or whisky instead. But, the "value transformation" which was the aim of this new settlement, where is it to be found here? Spiritual values are absent, or at least very little felt.

The material situation of the village is good, one of the farmers told me. Of course, some of them are poor, but so are some in any village anywhere. But there are also rich ones, many of them; and some had started with their bare hands. It is the spiritual situation that is not satisfactory. Any matter that cannot be turned to monetary profit is a waste of time for the Petach Tikva man. And the wild religious orthodoxy, own sister to the fanaticism of the Old City of Jerusalem, also prevents any spiritual life from developing. The atmosphere grows stifling, really stifling.

That is the first Palestinian antinomy: In order to reform economic life and conquer labor, settlement must be carried out by simple, uncouth men. But in order to bring about a "value transformation" and revive spiritual life, we need men who are educated and idealistic. Petach Tikvah succeeded economically precisely because its inhabitants are not idealists. But in the spiritual, intellectual sphere it cannot provide a model for the Jewish village. . . . What is the solution? [27]

The Fourth Period: 1914–1937

The First World War stopped the process of rapid growth and endangered the very existence of the colony. The first blow was to the citrus branch, which depended on European markets. When these closed, an immediate dearth of revolving capital was felt and regular work in the groves had to be discontinued. Money from abroad, efficiently administered, enabled the colony to get along in the first war years and the farmers to cultivate the groves. When, however, for a period of over two years Petach Tikva was on the battlefront, many houses were destroyed, the inhabitants left, and the groves were seriously damaged. The colonies in the Galilee were less damaged at that time, not only because they were in the line of fire for a shorter time, but mainly because their economy was geared to the satisfaction of internal needs and not to outside markets.

That Petach Tikva escaped destruction was largely due to the organizational ability of its inhabitants. Intensive organization during those difficult years was characteristic of all groups in Petach Tikva. The council organized a series of activities: concentration of all foodstuffs and their distribution among the inhabitants; administration of money contributions from abroad for the cultivation of the citrus plantations, by using it as payment for work done; and the issue of local paper money. The last two measures enabled the colony's inner economy to function and ensured the distribution of money to farmers and laborers alike. This energetic and constructive activity of the farmers' council was only geared to emergency needs, and it stopped with the end of the hostilities. The laborers' commune, on the other hand, which was organized during the emergency, continued to function after the war and developed new patterns in the Jewish settlement work.

The reconstruction of Petach Tikva after the war was swift. In 1923 the population reached five thousand, i.e., tripled in thirteen years. The area of land had expanded to six thousand acres. A new wave of prosperity came to the country, and the citrus-growing branch flourished again. In the period 1924–29,

Table 7. Distribution of plantations by size, 1915 and 1932 (in per cent) *

Size	1915	1932
Big grove (12.5 or more acres)	15.0	18.5
Middle-sized grove (5–12 acres)	62.0	40.5
Small grove (1–4 acres)	23.0	41.0

* The distribution of area of plantations by size is given in Table 8. Compiled from *Bustanai*, Nos. 43, 44 (1933).

1,250 acres of citrus groves were planted, as much as the total before the outbreak of the war. A census carried out in 1932 by the Farmers' Federation in Palestine supplied data about the Petach Tikva plantations. The total area of fruit-bearing plantations was 291 acres. There were also 540 acres of young plantations. Table 7 reflects the distribution of the fruit-bearing plantations.

The figures show that the relative number of big groves remained more or less constant, while the number of middle-sized ones dropped; the number of small and very small ones grew. The relative decline in the middle-sized groves is apparently the result of the economic hardships of the veteran settlers during the years of the First World War, while the rise in the number of small groves seems to be the result of the appearance of new groups in the citrus branch in the years of prosperity after the end of the war. The owners of these small groves were professionals in the branch but had no land of their own; they were skilled workers, as well as small investors—merchants, clerical workers, etc. A certain amount of land distribution following the death of original owners may also have played a role. But as we shall see below, the problem of fragmentation was not severe.

In spite of the relative decline in the number of the middle-sized plantations, their area is still identical with that of the big ones, and together they account for 83 per cent of the total.

Table 8. Distribution of area of plantations, 1932 (in per cent)

Size	Owned by a single person	Owned by more than one person
Big plantations (12.5 or more acres)	29.2	41.5
Middle-sized plantations (5–12 acres)	49.6	41.5
Small plantations (1–4 acres)	21.2	17.0

Compiled from *Bustanai*, Nos. 43, 44 (1933).

Thus, the problem of fragmentation does not arise, although in the thirties many plantations were already inherited: Petach Tikva had at that time completed its fifth decade of existence, and many of the first settlers were dead or unfit for work.

Some 27 per cent of the area about which data can be obtained (2,542 acres) was owned by more than one person: heirs, partners, etc. A table which excludes groves that have more than one owner reflects a different picture of the distribution of groves by size: the percentage of big plantations declines, and that of middle-sized and small ones rises. These data point to one of the most frequently used methods of solving the problem of land inheritance: the heirs are partners in cultivation and divide the profits among themselves, but not the property itself.

This problem of fragmentation through inheritance was one to which the Petach Tikva settlers devoted much thought in the twenties and thirties, years when fruit cultivation was at its peak and the number of adult sons was already large. There was a body of opinion against land division, owing to this fear of fragmentation; but no collective, binding decisions were reached because of the individualistic nature of the social framework. There are no precise data about the way each family solved the problem, but the rich documentary material from this period points to two solutions: common cultivation of groves by the several partners owning them; or the investment of profits from the grove, during the father's life, in planting new areas for the younger sons. The constant expansion of plantation area made it possible to ensure the economic future of the sons. There was also organized settlement of settlers' sons in new localities, through a youngsters' organization called the Sons of Benjamin. (In this way Nathanya, Herzelia, and Raanana were founded). At the same time, a certain percentage of the children left the agricultural branch altogether, and some even left the colony, without receiving a portion of the property.

At the beginning of the twenties an intensive process of urbanization began in Petach Tikva, accelerating in the thirties until the locality was granted urban status in 1937. The same economic openness that had helped to develop agriculture now became the basis for abandoning agriculture completely. The process itself was gradual but continuous. As early as 1900, visitors described the small-town character of Petach Tikva:

> Petach Tikva is a colony, that is, a village that looks like a town or a town that looks like a village. Upon coming to the colony, we see a village with large fields and vineyards, good, fat soil that will be fertile if tilled properly. The vineyards are full of vines that would yield good fruit to a good landlord. But when you enter the place, there you see a town, with all a town's characteristics . . . full of trade, buying and selling.[28]

And yet as late as the end of the thirties, when the locality got municipal status, the area under citrus cultivation was still growing.

This continuity is also noticeable in the existence of many occupations that are quasi-agricultural and quasi-urban (see Table 9):

The number of farmers who live on agricultural profits alone is very small; I doubt whether fifteen out of every hundred plantation owners live on profits from sale of fruit alone. Many of the inhabitants keep horses as a means of livelihood. The horses are used for plowing and for transportation of fruit and various items needed for the cultivation of plantations. Some farmers are merchants or commission agents for various things such as chemical fertilizers, etc.

Table 9. Occupational distribution of the Petach Tikva population, 1930

Total number of inhabitants	7,000
Total of farmers	400
Resident farmers	250
Craftsmen	330
Clerical workers (male and female)	84
Shopkeepers and merchants	270
Drivers and carters	86
Jewish workers	1,600
Physicians, nurses, chemists	32
Other professionals	12
Policemen and watchmen	30
Religious functionaries	12
Teachers and kindergarten teachers	54

Source: Bustanai, No. 38, p. 17–18 (1930).

Some of the plantation owners are shopkeepers—there are more than sixty shops in Petach Tikva. Some trade in oranges, citrons, almonds, barley; some are millers, etc. Not a few are work managers on other people's plantations. Some live by renting rooms and flats; others by selling land for building; there are building contractors, etc.[29]

The compiler of these demographic data, published in an article in 1930, summarizes: "In Petach Tikva and its environs, there are 10,000 breadwinners (including Petach Tikva suburbs and the Arabs who work in the colony and live in their own villages) on an area of 8,000 acres. Thus, each person gets his livelihood out of an area of less than one acre."

In those years intensive industrialization began in Petach Tikva. In 1934 an industrial district was zoned and many workshops of various kinds moved in: building and heavy industry materials, food, chemical enterprises, etc. The thirties were a period of intensive urbanization in another way as well: a con-

Table 10. Growth and development of Petach Tikva, 1887–1932

Year	No. of inhabitants	Area (in acres)	Cultivated area (in acres)	Area under citrus trees (in acres)	No. of houses
1887 *	319	3,571	1,500	—	47
1900 †	818	3,310		150 (total of orchards–1,575)	76
1922 ‡	3370			1,460	
1930 §	7000	6,738	3,825	2,650	650
1937 ‖	20000			3,750	
			field-crop area: (in acres) #		
1886			1,225		
1891			1,800		
1914			1,250		
1920			875		

* *Bustanai:* Vol. 10, no. 27.

† *Ibid.*, Vol. 2, no. 38, vol. 4, no. 37.

‡ *Ibid.*, Vol. 6, no. 31.

§ *Ibid.*, Vol. 4, no. 37, vol. 2, no. 38.

‖ *Ibid.*, Vol. 8, no. 32, A. Tropa, *Foundations on the Occasion of Petach Tikva's Seventieth Anniversary*, in Hebrew (Petach Tikva: A. Tropa, 1948), p. 70.

#*Ibid.*, pp. 22–23.

Note: Discrepancies are due to the fact that the several sources drawn upon gave variant information.

siderable growth in population and in density. The increasing density of population and the consequent rise in real estate prices led to multistoried buildings, increasing the colony's urban character.

In Table 10 is reflected the process of transition from mixed agriculture based largely on field crops to a fruit-growing colony that grows into a town. The peak period of field crop farming is the nineties. In 1891 there were 1,800 acres under field crops. When vineyards began to be intensively planted at the Baron's instigation, the transition began, and in 1914 there were only 1,250 acres under field crops; in 1920, 875. At the same time, the fruit orchards grew in size. In 1900 there were 1,575 acres of plantations, but citrus accounted then only for 150. By 1922 citrus had become the main agricultural branch, with 1,460 acres, and by the time Petach Tikva became a town, there were 3,750 acres of citrus fruit. The growth of citrus plantations was most intensive during the twenties. These are the peak years of Petach Tikva as a fruit-growing colony. In the thirties the tempo of citrus plantation growth slowed down. In 1932, 2,905 acres of fruit-bearing plantations were registered; in the 1936 census the area had grown by 23 per cent. This is the lowest rate of growth as compared with all other Judaean colonies, which during the same period had grown by 90 to 150 per cent. In 1932 the Petach Tikva groves accounted for 28 per cent of the whole citrus area, but in 1936 they accounted for only 19 per cent. The transformation of Petach Tikva into a town is illustrated even better in Plate I and Map 6, which show the size, layout, and land usage of the colony in 1939 and 1951 respectively.[30]

The process of urbanization provided a solution for the socio-cultural stagnation. A first breach in the wall of traditional conservatism was driven by means of population growth, in the struggle between the old and young people, between the laborers and the farmers. With the onset of the urbanization process this wall crumbled completely; Petach Tikva became a town with a modern way of life. This provided the solution for those individuals who were irked by the traditional limitations, but it was by no means an innovation or a cultural revival. Petach Tikva may have solved its own cultural problems but created no new social patterns or values. It is no accident that very few of the ideological and political élite of the country come from Petach Tikva. Moreover, to this very day Tel Aviv, which lies at

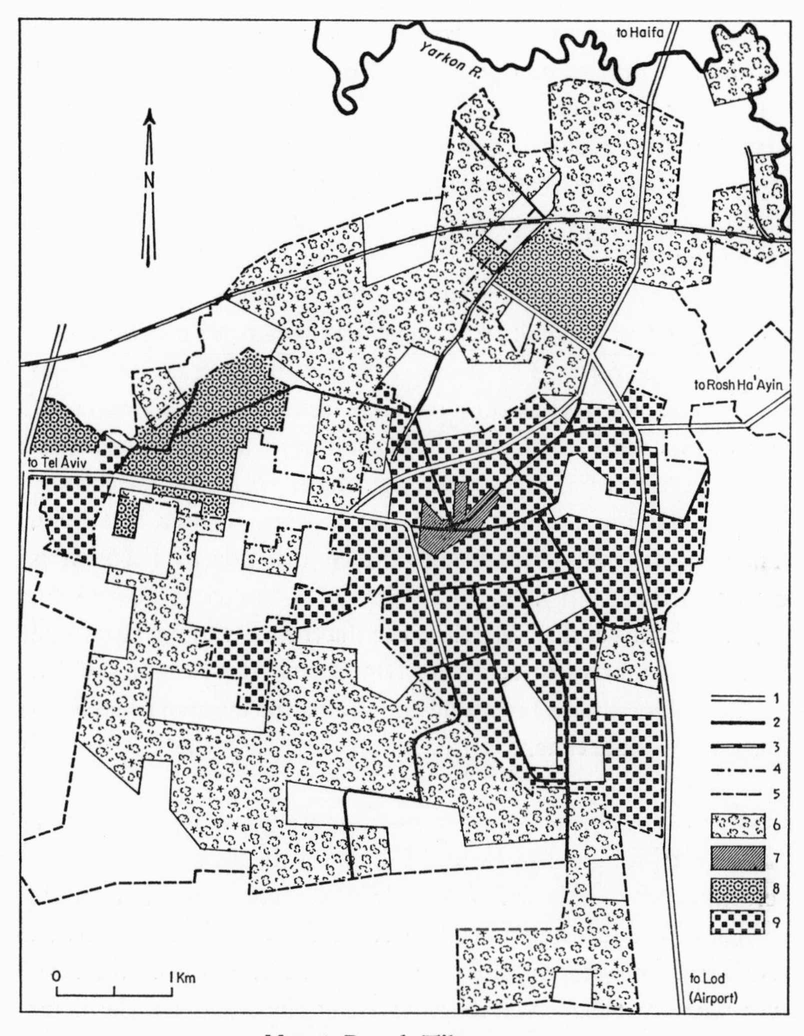

Map 6. Petach Tikva, 1951
1. main road, 2. road, 3. railway, 4. municipal boundary, 5. local planning boundary, 6. orchards, 7. commerce, 8. industry, 9. residential area.

no great distance, serves as the cultural and recreational center for the Petach Tikva inhabitants.

A colony that became a town is one of the three types into which veteran Israeli villages can be divided. The other two types are colonies that developed but remained agricultural in character, and stagnant colonies. It is to be assumed that the main factors that determine the direction of development are type of soil, water resources, contacts with internal and external markets, and the developmental potential of various types of farming: fruit farming requires intensive work while yielding high profits, field crops require extensive cultivation but have low profit margins.

Petach Tikva was blessed by good conditions in all these respects. It had possibilities of development for both agricultural and urban patterns. The decisive factors in the transition from the agricultural to the urban pattern were the limited amount of land, which thus became expensive and more suited to industrial than to agricultural investment, and the possibility of becoming integrated in the urban economy of the nearby center, Tel Aviv. The impact of this proximity to the conurbation of Tel Aviv is clearly seen in Map 7.

Perhaps more than all this, the factor of timing was here decisive: the crisis in fruit farming during the Second World War combined with favorable conditions for industrial development at the same time.

Summary

The establishment of Petach Tikva was due to a movement of reform among Orthodox Jews. Living in Jerusalem, they strove to change the Jew into a productive person who lives by his own work and does not subsist on charity. Redemption of the land was seen by them as a religious obligation, and the Jew, by becoming a farmer, would be a better man whose everyday activities in agriculture would be the fulfilment of this great obligation. The founding fathers, far from wishing to dispense with religious orthodoxy, strove to strengthen it. This did not minimize at all the innovation which their conception entailed. The focus of the human image was transferred from spiritual occupations to concrete activity in the settlement of the land. The innovation was the outcome of an inner growth of religious values and not the result of exchanging them for others.

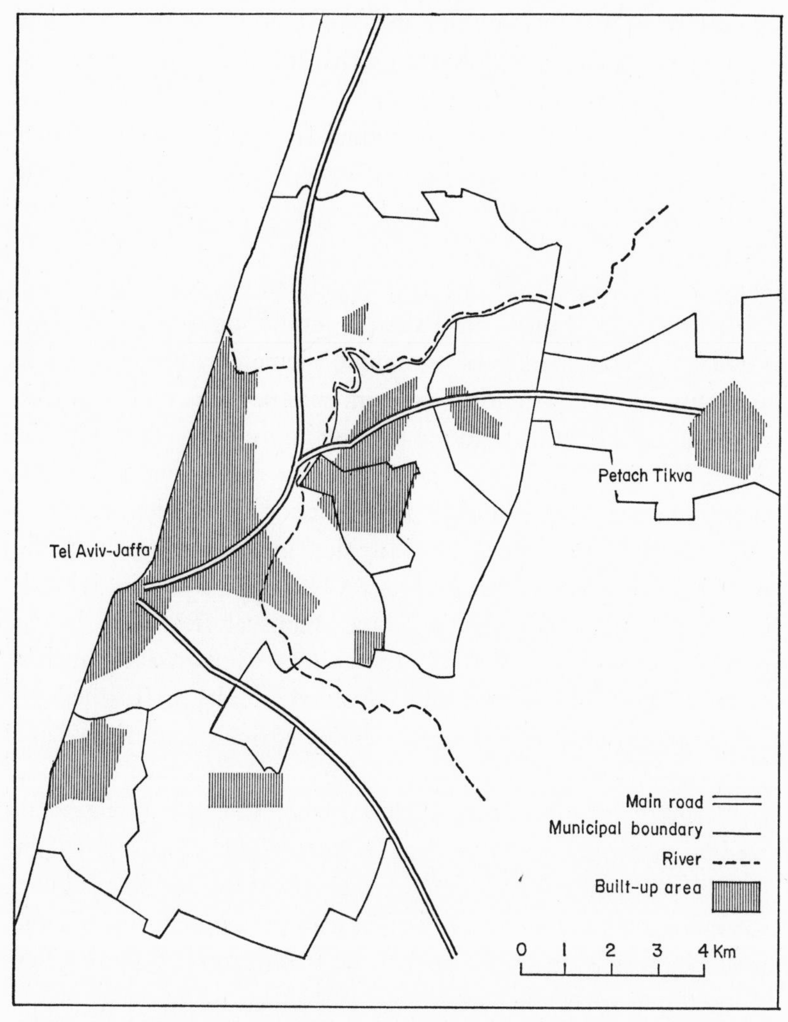

Map 7. Petach Tikva in relation to metropolitan area of Tel Aviv, 1944

The value innovation was thus the source of the economic innovation, and the first Israeli agricultural village was established in 1878 in the center of the swampland of the Sharon Valley in order to realize the ideal of land reclamation. The first experiment failed and the settlers left. But after a year they returned, determined not to abandon it again. But it was precisely the belief of these pioneers that each individual family's ability to maintain itself on the land by its own work had historic implications for the nation that ultimately led to a transformation of the ideal itself.

The value transformation stemmed from the central concept of "living by the sweat of one's brow." The applying of the principle of living off the land brought economic criteria into prominence. Economic activity became the yardstick for the evaluation of the idea itself; the settlers had to prove that Jewish agricultural settlement in Palestine could be self-supporting. Hence the ascendancy of economic elements over ideological elements in the farmer's role conception. The individualistic economic motivation became the main driving power in the colony's history. The demographic pattern—large families working on the family farm—strengthened this tendency.

The transformation of the idea that inspired the colony did not hinder its further development. On the contrary, individualistic economic motivation produced unusual flexibility in the sphere of economic behavior: the colony went over from field crops to fruit orchards, from extensive agriculture designed for the supply of internal needs to intensive agriculture designed as a commercial branch. The commercialization of agriculture and the large-scale employment of hired labor was in complete contradiction to the original ideal of "growing bread out of the soil." But the transition to fruit farming was the turning point in the colony's development: it put its economy on a firm basis and made intensive farming possible, and ultimately it determined the colony's social character.

At first the economic flexibility was not paralleled in the sphere of social behavior. The colony's public institutions underwent no change for three decades, and the founders jealously defended the traditional way of life taken over in its entirety from the Old City of Jerusalem and from European ghettos. But the colony grew and changed. The transition to fruit farming brought new men to the colony, among them workers from the Second Aliya who were radically different from the old-timers; economic growth and change in type of farming led to a differ-

II. Aerial view of Ein Harod—1946. The picture shows the kibbutz compound. In the center of the picture are the communal dining room and other public buildings. At the left and right front are located the members' rooms. At the back one can see the workshops, poultry batteries, a stable, and cowsheds.

I. Aerial view of Petach Tikva—1939. The picture shows the center of the town. In the twenties most of this area was the site of citrus plantations.

entiation in the occupational structure and in the structure of the population. The homogeneity which was the basis of integration in the first stages was thus undermined. Two severe conflicts divided the colony. But the fact that these two conflicts caused identification along two different lines of division (farmers versus laborers, and old versus young) prevented a total split and led to a new integration. The demographic change in the colony's population, which resulted from economic development, ultimately breached the wall of social stagnation. This in turn enabled Petach Tikva to develop in the direction of full urbanization.

Thus Petach Tikva underwent another change, from village to town. Here too the motive power was economic. This motivation, entirely divorced from the original values, led to an increasing rejection of the agricultural pattern and to the development of urban types of activity which opened new economic possibilities. The unusual economic flexibility—the result of value transformation—was the cause of both the intensification of agriculture and the abandonment of it at a later stage.

Petach Takva is a success story. It is now one of the biggest towns in Israel; it pioneered the cultivation of citrus fruit, which is today one of the main exports of Israel; later it was in the forefront of industrialization; and it absorbed a great number of immigrants. In some respects, mainly the absorption of immigrants, Petach Tikva was more capable of responding to new social needs than the other two types of agricultural settlement in Israel. Paradoxically, the abandonment of the original ideal created a better capacity to respond to new national needs. But because of this value transformation, it did not initiate any new type of economic organization or novel social ideals and style of life, as the moshav and the kibbutz did. The only genuine contribution of Petach Tikva in this respect was its founders leaving the Old City of Jerusalem and building the first Jewish agricultural settlement in the heart of the wilderness.

3

A Kibbutz: Ein Harod

Ein Harod was not the first collective settlement of its kind in the sense that Petach Tikva was the first moshava, and Nahalal the first moshav. However, while it had been preceded by the small kvutza, it created a unique pattern and gave rise to a settlement movement of paramount importance. Without any hesitation, we rank the settlers of Ein Harod among the founding fathers.

The history of Ein Harod is presented in three parts: the background, up to the actual establishment of the settlement in 1921; the period of social unrest and crisis, 1921–1926; and the period of sustained growth and development from then till 1948. The dividing line of 1926 is mainly a social and institutional one, marking the crystallization of a viable community and structure; in the economic field the turning undoubtedly occurred in 1931, with the transfer to the permanent site and the allocation of new land and capital. This transfer signaled a change from almost complete economic dependence on the settlement authorities and a chronic state of scarce resources, to autonomy and economic drive and confidence. The chronological division adopted is thus somewhat arbitrary, Ein Harod not having been considerate enough to develop in a balanced way.

The First Period—Forefathers and Founders (till 1921)

The initial venture in collective settlement in Palestine was the small kvutza (group). The founders of the settlements of the kvutza type (the first of which was Degania, established in

1910) were pioneers of the Second Aliya, who had arrived in the country in small groups before the First World War. Their first steps in the new country were extremely difficult, and they had to face the physical hardships, sickness, and deprivation of their new and stark environment, compounded by the adjustment to unfamiliar agricultural work.

The newcomers began by working in the veteran colonies: they had then no clear aim of founding their own settlements, but hoped for a part in the general colonizing effort, and in regeneration through agricultural labor as such. In the moshavoth, however, they faced the rude awakening that awaited the pioneers of the Second Aliya: they were compelled to fight for the right to work in competition with low-wage Arab laborers and against a hostile social environment. In the study of Petach Tikva we saw this situation mostly through the eyes of the farm owner who, perhaps not unnaturally, found it difficult to accept on his terms the proud, dedicated, but unskilled pioneer. From the point of view of this pioneer, however, the supervision of foremen and gang leaders, who treated him as an inferior and as a liability, deeply offended his dignity and self-esteem, and subverted his basic national and social image. The stress of adaptation was made still harder during that period by the lack of any organization or framework to offer support and help; each individual had to struggle on his own.

This reality of a pioneering intelligentsia becoming a landless rural proletariat aroused the desire to establish independent workers' settlements, in which it would be possible to work without supervision and to create a different kind of society. Groups of workers organized, accordingly, with the purpose of obtaining land and funds from the Zionist institutions. Their claims met with the approval of Arthur Ruppin, then head of the Palestine Office of the Zionist movement, who valued the creative force and national idealism that motivated the young pioneers and who took up their struggle within movement institutions. With his help, the new venture of settlers colonizing national land with national funds was launched; this resolution was to become the main principle of Zionist settlement in years to come.

The independent settlements established by the Second Aliya workers initially assumed several forms and underwent different experiments, including the establishment of training farms under the management of agronomists; finally, however, they crystallized into the small kvutza. The small kvutza was based on a tightly knit little group. It copied from its predecessors the single-crop plantation, which became its basis of existence. Otherwise, though, cultivation was completely collective and the way of life communal, including consumption and socialization. The village became a highly egalitarian society, as well as an intensive and direct democracy, with the settlers performing all the necessary economic, public, and service functions themselves, in rotation. In addition, the small kvutza assumed the task of training inexperienced workers (its own and others) in agriculture, through utilizing the potential of its collective production system.

During the First World War and immediately afterwards, the small *kvutzot* (plural of kvutza) faced a severe crisis that had its roots in several factors: the settlements did not achieve economic self-sufficiency and were permanently in debt; the small size of the community and the limited scope of its farm enterprise impeded not only development but also absorption of new members, and the kvutza could not cope with the increasing immigration to the country; the social smallness of the kvutza intensified the relations between the members and increased their interdependence to a point where the individual was left almost entirely without privacy, and this created severe tensions; and finally, the separation, geographical and social, in a narrow and isolated framework produced a sense of withdrawal from active participation in general social and cultural life.

The groups were thus unable to support themselves within their narrow economic and social limits, or to absorb and provide employment for new immigrants at a time when this was a vital function. This crisis brought about the disintegration of some of the kvutzot and a high level of desertion in those that continued to exist; and it inevitably caused changes in the structure and orientation of the groups.

The changes assumed different forms: some kvutzot retained their original pattern and, strengthened by a process of selection, based themselves on an élitist consciousness and on the social and ideological bonds uniting the members; many of the leavers abandoned the collective pattern of settlement and chose the moshav instead; others wanted to retain the essential communal nature and the pattern of organization of the kvutza, but to broaden its social and economic scope and to rationalize its methods of cultivation and division of labor. This stream, combining innovation with continuity, was the main basis of Ein Harod; and while the majority of the actual founders who first settled and organized the site belonged, as we shall see, to a different group with a different background, it was here that the ideological foundation was laid.

The chief spokesman for the large kvutza (later large kibbutz), evolved in reaction to some of the features of the earlier form, was Shlomo Lavee, a veteran worker of the Second Aliya and a member of Kvutzat Kinneret. He was deeply committed to the basic values of the collective and communal way of life, and in his new ideas he incorporated those tenets of the kvutza relating to organization of production, consumption, socialization, and community life. However, in a series of publications he put forward a drastically different social and economic guiding image, which, when examined carefully, presents a seven-pronged ideological program:

1. *An economy based on size and diversification.* The small kvutza, Lavee pointed out, had a limited capacity for both economic and demographic growth. This limitation, inherent in a monoculture of modest proportions, would in the large kvutza be drastically changed by means of an *integrated and diversified economy*, which would develop agriculture, handicrafts, and industry side by side, and so create a large potential for employment and absorb a larger number of immigrants than would be possible in a pure farm economy. The new concept would also

help overcome the scarcity of agricultural land, which constituted an absolute obstacle to settlement, while internally it would make possible a more rational utilization of manpower and resources and thus solve the chronic economic problems that had been the bane of the kvutza:

> In our conception agriculture is perceived as a basis for industry, construction, and public works. The absorption potential of the country can be realized only if we strive for maximum development of every square kilometer, cultivate and utilize it to the full, incorporating all the sectors of the country's economy. This involves development of diversified economic areas, diversified farming which comprises agriculture and industry.
>
> Industry in the kibbutz enables the reduction of production costs by providing maximum employment for the labor force of the settlement all through the year and the reduction of the cost of the everyday necessities of the workers in industry . . . from the agricultural point of view, the addition of branches, introducing diversity into the economy and increasing the output of the workers, contributes to the strengthening of the agricultural unit. . . . We have to devise methods of work ensuring high productivity in order to achieve a decent and civilized standard of life.[1]

2. *Agricultural autarchy and mixed farming.* The agricultural sector of the combined or integrated economy would be based not on monoculture but on mixed farming. This concept had its roots in the crisis among the citrus-growing colonies during the First World War, when the economy of these colonies, based mainly on export to foreign markets, had suffered a severe slump. The founders of the kibbutz sought to eliminate this dependence and prevent the possibility of future crises. The literature preceding the foundation of Ein Harod stresses the importance of autarchy and the need to eliminate the farm's dependence on the success of any one particular branch.

However, over and above the purely economic aspect of the concept of autarchy, mixed farming with different agricultural branches was also well adapted to one of the guiding principles of the kibbutz: self-labor. Such branch structure implied an even work curve and would guarantee employment for the majority of members during a longer period of the year than would a monocultural farm. By the same token, it would prevent any seasonal rush, with its demand for large numbers of additional manpower to carry out the work of fruit picking and packing. The concept of autarchy and mixed farming, therefore, encouraged the idea of work carried out exclusively by the members of the settlement without the help of hired labor.

3. *Market participation.* In sharp contrast to the moshav, which, as we shall see, interpreted the concept of autarchy as a minimal consumer and producer participation in the goods market, here the segregation and autonomy was to be much narrower: the kibbutz would be integrated with the economy at large—at least as marketer—through its agricultural and industrial products. A similar difference from the moshav would apply also to the labor market; for while the kibbutz, like the moshav, rejected the hiring of labor, it emphasized the participation of its members in various outside economic enterprises.

4. *Orientation towards expansion and growth.* Intimately related with participation in the modern markets and drawing upon the immigrant-absorption ideology, was the orientation towards expansion and growth. In other words, the kibbutz would not only start on a broad and diversified social and economic basis, but would continue to develop in both areas, emphasizing sustained, comprehensive, and varied growth as a basic value. Thus, while the agrarian system of the kibbutz was to be circumscribed by the ideological limitations relating to branch structure and to source and division of labor, within these bounds there would be an overwhelming orientation towards a dynamic economy, with a consequent rational emphasis on the productive process and a drive towards constant experimentation and innovation. *At the basis of the kibbutz—in sharp contrast to the kvutza—there lay a fundamental entrepreneurial image, embodying sustained initiative and absorption of change, as well as constant experimentation, and allowing it to give scope to the "large business" potential embodied in its projected structure:*

The agricultural unit must be divided according to specialization and not according to types of work, in order to increase production. This division strengthens the tie between the worker and his work. Every specialized branch will be directed by a member-instructor, and these instructors together will form the council of specialists who will discuss matters directly related to the economic aspect of the settlement.[2]

Or:

The large agricultural unit affords the advantage of making use of machines, without which production cannot increase . . . work becomes easier, more pleasant and more rational. It requires the use of the worker's brain, mind, and experience rather than the use of his muscles.[3]

5. *The open-door policy*. As envisioned, the creation of the large kvutza would hinge on an open-door policy. While the kvutza followed the élitist image, regarding itself as a community of chosen people admitting only those preselected on the basis of a common ideology, the kibbutz would initially require only the basic acceptance of the values of self-labor and collective life; and it would itself further socialize the heterogeneous candidates.

The small kvutzot, being restricted and closed, were incapable of inducing movement in the life outside of them. . . . Is not the lack of economically gainful settlement the main, indeed, the sole cause of unemployment? The most radical cure for this is doubtless the establishment of this type of settlement, if we succeed in carrying out this aim. Such settlement provides an appropriate solution for the needs of increasing immigration, the needs of the immigrants themselves and the problem of underemployment, which is the direct result of the lack of creative work and the lack of a sound economic base for our activities in Palestine. Our task is to realize this plan, to establish the large settlement which will enable the absorption of immigrants and the foundation of the Jewish national state.[4]

6. *Social and individual fulfilment*. Perceptible differences between the ideology of the small and the large kvutza were also

found in the image of the "good village society." It is true that both groups had inherited a somewhat rigid social perception and ideology from the socialist movements in Europe and particularly the various revolutionary movements of Russia. They therefore regarded society as composed mainly of two distinct strata, with a somewhat small and loosely defined middle class and intelligentsia. At the top of the social pyramid we find the rich and powerful, who are the exploiters of society and enjoy all benefits of life, while at its base are the masses, peasants and workers, ceaselessly toiling and receiving only meager rewards for their labor. The intelligentsia, situated somewhere between the two, join forces with one or with the other of the main social strata. In the cultural sphere, only the members of the upper and middle strata are given the opportunity to develop their artistic and intellectual talents and are taught to appreciate cultural and aesthetic values; the masses, by contrast, are never given such an opportunity, do not develop their mental resources, and become boorish and unresponsive to cultural and aesthetic values.

The perception of the class society, common to most socialists of the period, was tinged also with a special desire to reform the traditional socioeconomic structure of the Jews in the Gola (Diaspora). This structure, particularly in Eastern Europe, was characterized by overrepresentation in secondary and tertiary occupations; by a preponderance of petty traders and middlemen; and an almost complete absence of workers and peasants. One of the essential aims of the reformers—of whom the members of the Second Aliya were a part—was to turn this inverted pyramid and to redress the balance by channeling the young generation to productive occupations in general, and to manual labor and farming in particular.

This process of productivization, which had been begun by the First Aliya and even earlier, was now found to be reversed. Although the range of occupations of the Jewish inhabitants of Palestine was much wider than it had been in the Diaspora, and although a rich and powerful class could scarcely be said to have emerged, various symptoms indicated a relapse into the traditional structure and the weakening of the ideals of the first Zionist pioneers. In the colonies, a planter stratum had thus developed an urban orientation and abandoned the ideal of agricultural work to be carried out by the colonists themselves. In the towns, the way of life and the occupations of the urban middle class reminded the people of the Second Aliya of the very things they had rebelled against in the Gola.

Both the founders of the kvutza and those planning to create the kibbutz envisioned the establishment of a completely new way of life and society that would eliminate class differences and create a bulwark against further deterioration in the values of productivity and equality.

In principle, the realization of individual potentialities was a cardinal virtue for the kvutza too. Its "good man" was one devoted to the cultivation of the soil, characterized by austerity, simplicity, and honesty and free from the hypocrisy and opportunism of a materialistic society; at the same time, though, he should realize the humanist ideal of a man living a full life, rich in social, cultural, and emotional experience and fulfilment. The theorists of the large kibbutz, however, maintained that the conditions of life in the small kvutza made the achievement of this aim impossible; it required a large and complex structure, abounding in variety and opportunities for individual development.

We have never advocated the establishment of a kind of mechanical entity, neither have we demanded asceticism or charity. We aspired to something greater, richer, and more expansive. Our aspirations were directed to uniting the free forces, the free wills, and strivings into a combined effort of people who knew how to attach themselves to the Great Power that has created the world wisely and according to plan, and not by "stichia" [emotional drive].

Self-denial is not required. On the contrary, all needs must be fulfilled and nothing must be renounced. A large society creates life, joy, and movement; these afford satisfaction to the individual and he is ready to give up things without even noticing it. The situation is completely different in a restricted society of a few chosen people, in which boredom prevails, people grow tired of one another until a man wishes to escape even from himself. This is the root of the disintegration of the small kvutza which occurs from time to time.[5]

Socially, therefore, the concept of the integrated economy made possible, in their opinion, the creation of a differentiated and pluralistic society, with a high level of culture, embodying the principles of equality and social justice. In other words, differentiation of roles was to be not only an economic necessity but a social asset, and it would assure individual rewards and motivation in a nonstratified structure. Thus, the exponents of the kibbutz idea envisaged a differentiated community including a variety of spheres of activity and encompassing not only agricultural roles but also roles characterizing urban life: industry, handicrafts, and public work; and this orientation toward differentiation and specialization was well adapted to the emphasis on rationality and development.

The innovation was in the attempt to combine and adjust between social and economic modernization on the one hand, and the values of collectivism and equality on the other, an attempt calculated to eradicate the significance of general differences in class and education, and to allow the manual worker to develop his talents and to enjoy the benefits of a rich social and cultural life. The alternative to urban life did not lie here, as in the kvutza (and the moshav), in the creation of a wholly distinct rural culture, but rather in the provision of some of the urban characteristics on different terms. "We are creating a society," said I. Tabenkin, one of the founders, "which is neither a town nor a village, but a large and growing collective workers' settlement. This settlement will be built in rural surroundings; it will have fields, crafts and industries, educational institutions and cultural activities. It will be founded on the ideals of friendship, collectivity, and joy in creative work."[6]

Of course, in reality the range of differentiation and specialization was of necessity comparatively small; but much of the history of Ein Harod can be understood in terms of the attempt to realize the basic values of collectivism and equality under the conditions of an expanding population and an economy growing in volume, efficiency, and complexity.

7. *Primacy of production in place of asceticism.* Another point on which the program of the large kvutza differed from that of its predecessor was that it did not preach the ideology of secular asceticism, austerity for its own sake and as a value in itself. Here limitations on consumption were essentially a necessity deriving from conditions of economic underdevelopment and from the primacy of saving and investment. However, the kibbutz considered the individual's desire to improve his material standard of life to be legitimate and stressed the achievement motive in this respect, both with regard to the individual and with regard to the community as a whole. The kibbutz was thus ideologically geared to progress and development not only in production, but also to a large extent in consumption, providing for a legitimate desire for a rising—though equal—standard of living and services; and this orientation reinforced the impetus toward constant innovations and improvements.

All in all then, while the kibbutz retained from the kvutza the collective unit of production, consumption, and socialization, the egalitarian way of life, the emphasis on public land and financing, and an image of a democratic and autonomous community, it made these values into something essentially new and different. Like the founders of the kvutza, the exponents of the kibbutz articulated their idea with great clarity and showed total commitment to it. These qualities can indeed be traced to the common background of the two groups, in particular, to the tradition of Russian Jewish intellectualism and the Russian socialist movement, with their constant preoccupation with ideas and their belief in the force of ideology.

This, then, was the major strand in the guiding image of Ein Harod, ideologically sustained and socially supported by settlers of the Second Aliya—mostly former members of the small kvutza. The second major strand, and very distinct from the former, was that of the Labor Battalion—Gdud Haavoda.

The Labor Battalion was the characteristic achievement of

the Third Aliya. It was founded by young men and women who had grown up during the period of revolutionary upheaval in Russia. They had absorbed the socialist-revolutionary ideals of their environment, but after the revolution had become a reality, they realized that as Jewish nationalists they would not be able to preserve their identity in the new social order. The situation in which they found themselves prompted them to emigrate to Palestine with the aim of realizing their socialist ideals in the developing Jewish community, which was at that period in the initial stages of its growth.

Having received their education in the "University of the Russian Revolution," they developed a radical orientation toward social change, the realization of which required revolutionary means and demanded the total commitment of the individual and the negation of personal interests. The achievement of their conception of social change required the creation of an efficient organization that would enable them to establish the political hegemony of the working class, which they regarded as the vanguard of the new social order based on the principles of equality and social progress.[7]

These pioneers of the Third Aliya wanted to participate in the national revival not necessarily by founding new agricultural settlements, but by the implementation of development projects in general, mainly construction. Their concept of the Labor Battalion as an army of workers, conquering strategic positions in the country's economy for the working class, derived directly from the Russian Revolution. This concept was connected with the acceptance of a centralized authority, readiness to perform difficult tasks, political alertness, and willingness to move from place to place according to demand and dictate. As socialists, they accepted the idea of cooperation, but its interpretation was stricter, influenced by their role as an army of conquest. Thus, in their organization they preserved the principle of complete equality of consumption in and among all units by a common treasury, designed to maintain equal living standards among the various component companies of the Battalion. Life in the Battalion was, however, regarded by its leaders and by the rank and file as a passing phase, lasting only as long as necessary to fulfil the task of labor conquest. In the future, they envisaged a nationwide commune of the Jewish workers in Palestine based on the principles of collectivity and equality.

The members of the Battalion were young people in their twenties, without family obligations and without any firm roots in their new environment, thus constituting a suitable human element for the realization of pioneering goals. The Battalion was organized in the form of work groups attached to public-works projects of the mandatory government. Discipline was maintained by means of a centralized organization and by the authority of the leaders, who kept up a daily contact with the rank-and-file members.

The Histadruth,[8] the General Federation of Jewish Workers (founded in 1920), granted its support to the Labor Battalion, which it regarded as an important factor in the realization of one of its main goals, the penetration of Jewish workers into new occupational branches, and the provision of work for the new immigrants arriving in the country in growing numbers as a result of the Balfour Declaration.

These, then, were the two strands from which Ein Harod was to be fashioned.

The Second Period—Creation and Crisis (1921–1926)

Ein Harod's actual beginnings are to be found in a group of seventy-five members of the Labor Battalion, who constituted the advance party of a new settlement, and who settled on land just bought by the Keren Kayemeth near the village of Nuris in the Jezreel Valley. The original site was near the spring of Harod (*ein* means spring); but several weeks later another part of the work company arrived and pitched its camp on a nearby spot called Tel Yoseph, so initially the new venture included twin sites and was named Ein Harod–Tel Yoseph. Within a short period, the settlement was joined by a smaller group of Second Aliya members, who had developed the idea of the large kvutza and had been waiting for the opportunity to put it into practice, not judging themselves large enough before for this purpose.

As can be seen from the brief sketch of the background of the two elements, there were considerable differences between them. The minority could draw upon their previous experience as workers in the colonies and as members of kvutzot, and they regarded themselves as well informed on the problems of agricultural settlement and knowledgeable in collective living. The Battalion members, on the other hand, were new in the country, unfamiliar with its conditions, and lacking any previous socialization to life as settlers. Also, their experience in collective organization was limited to the common treasury, pooling wages of the members to cover the living expenses of everyone; the intricate problems involved in the management of a unit of production were unfamiliar to them. The Battalion thus provided a group of young, well-organized, and idealistic people who were willing to adopt the new idea of the large kvutza, but who were neither committed to it nor prepared for it.

The fundamental distinction between the majority and the minority, however, lay in their diametrically opposite conceptions of the place and significance of the kvutza in their lives and in national tasks. While both sections hoped that the joint endeavor would bridge the gap between them and create a common interest in the new settlement, the ideological differences—sharpened by actual difficulties—proved too strong. The veterans of the Second Aliyah were then completely oriented towards the settlement, which was their sole aim and purpose, and they regarded their alliance with the Battalion as an expedient towards the achievement of this aim. The leaders of the Battalion, on the other hand, regarded the settlement as a means of increasing the weight and influence of the Battalion. The important thing for them was the preservation and strengthening of the Battalion, not the success of the venture in agriculture. Their ultimate goal, it will be remembered, was the creation of a countrywide commune composed of Battalion work groups, which would gain dominance over the important areas of the country's economy; they were prepared to support the venture of Ein Harod only as long as this appeared to further their general aim.

But conditions of unemployment in Palestine now began to threaten the very existence of the Battalion elsewhere, and its leaders, in their anxiety to prevent disintegration, made use of the scanty funds that had been secured with great difficulty for Ein Harod from the Zionist settlement authorities, to refill the empty treasury of the Gdud. This was justified, in their eyes, on the basis of the principle of a common treasury; regarding the settlement as just another unit of the organization, the success of which seemed doubtful in any case, they saw no reason to endanger the survival of the Battalion for the sake of saving the settlement. The veterans, on their part, were convinced that the lack of economic independence of the kvutza completely eliminated any chances of its successful development. They strongly resented the daily interference of the Battalion's leaders in the affairs of the settlement, and particularly the use of its funds to finance the losses of the Battalion.[9]

The conflict became exceedingly bitter, and relations between the two groups deteriorated to a point where all cooperation was impossible; the two sectors turned into two factions, each striving to impede the efforts of the other. In this situation, the Labor Battalion members, constituting a majority and forming an integral part of a centralized organization, had a dominant position. The Second Aliya settlers, however, were able to gain the support of the Federation of Labor in this conflict. The crucial importance of the Federation, and the reason why the Ein Harod minority could influence its decisions, are documented in Chapter 6. Here it must only be mentioned that the Histadruth, which had previously recognized the Battalion, was now aware of the difference of interests between them and becoming suspicious of the Battalion's growing independence. It thus intervened through its Agricultural Center on behalf of the minority and swung the decision against their expulsion, in favor of a division, which was implemented a year and a half after establishment.[10]

The settlement, which had been founded on two sites with the aim of eventually establishing additional units as part of a planned comprehensive settlement according to the ideology of the large kvutza, was now divided between the two factions, each receiving one of the sites. The majority, comprising most of the Battalion members, received the site of Tel Yoseph, situated on a hill close to the original site of Ein Harod. The minority, comprising the veterans of the Second Aliya and a few Battalion members, received the site of Ein Harod itself, situated in the valley near the spring of Harod.

This division served to accentuate the dynamic orientation of the Second Aliya settlers, the exponents of the large kvutza ideology. For while the site of Tel Yoseph included the dairy and field-crop branches, which were of great actual economic value and considered the staples of collective settlement, the area of Ein Harod had all the experimental branches, like plantations, machinery, and workshops; these branches, though not yet profitable, were designed to form the basis of a future integrated economy and represented the values of expansion and innovation. The division left Ein Harod, which now becomes the sole subject of our story, with 105 members and 22 children. And while the separation from the Battalion enabled the settlement to achieve independence, its population and resources now fell below the minimum required for the realization of the ideal of the large kvutza.

The split in the kvutza, which occurred in the initial stage of settlement before the new enterprise had been given a chance of proving its potential for development and stability, was a great blow and left the pioneers of the settlement in a state of dejection. From the start they had encountered strong resistance to their venture on the part of the Zionist settlement authorities, and even though they had received some support from the workers' sector of the Yishuv, this was far from enthusiastic; it appeared as if all those who had opposed the venture or had doubted its success had been proved right. However, the exponents of the kvutza ideal had from the start been characterized by attributes that now enabled them to overcome the crisis: they had a definite idea of what they wanted to create and a drive to realize their idea; they had a strong urge towards innovation and a consciousness of their mission as the vanguard in the realization of a new social ideal; last but not least, they also had among them leaders and people gifted with initiative, experience, and organizational competence, capable of finding practical solutions to the problems at hand.

Thus, the fact that they had been left alone to continue their battle against adverse conditions did not shake their belief in the ultimate success of their venture, and they proceeded to seek an element that could be induced to accept their idea of the large kvutza and who would, by joining them, enable them to continue with their plan. They found a group of pioneers who had immigrated to Palestine in the Third Aliya and were engaged in ground-clearing work in the Jezreel Valley, not far from Ein Harod. The members of this group resembled the minority group at Ein Harod in their background and outlook, although they were on the average several years younger. They too were devoted to the idea of founding a settlement and experienced in agricultural work and in patterns of collective living. They had followed the development of Ein Harod closely and, though not previously committed to the idea of the large kvutza, were ready to endorse it. After due deliberations on both sides, the union of the two groups was made effective in 1926, only a few months after the split with the Battalion, and the united body was called Kibbutz Ein Harod. The new group—Chavurat Haemek (the Band of the Valley)—which was composed of several small units working in the area on various agricultural projects, still maintained two such nuclei outside the framework of the settlement. However, the stated aim of the new body was the union of these dispersed elements with the settlement as soon as economic conditions allowed it to absorb additional members.

At the time of the union, the number of settlers at Ein Harod belonging to both groups reached 173 and a few months later rose to 202. After the union, the settlement entered a phase of primary consolidation. The skilled manpower introduced into it made a good start in the main agricultural branches possible. Both groups agreed to the emphasis on the autonomy of the settlement and attached supreme importance to the creation of the necessary conditions for its successful development. The settlement was regarded by the members of Kibbutz Ein Harod as the central objective and the work groups only as a temporary arrangement; their members were expected to develop an attachment to the settlement. The similarity of background and orientation of the two elements and their previous practical experience in the country before settlement made the success of the union between them much more feasible than the previous attempt. The breathing space it secured, however, was short-lived.

The strengthening of the link between the settlement and the Histadruth, and the fact that the latter could no longer rely on the Battalion to perform the social function that it considered essential, caused it to request Kibbutz Ein Harod to substitute for the Battalion and send out a company of members to a distant site to carry out a public-works project under the direction of Solel Boneh, the Histadruth contracting concern.[11]

Since the prospects of consolidation of the settlement appeared much better now after the union with Chavurat Haemek, the majority of its members tended to accept the challenge presented by the Histadruth, claiming that the kibbutz could now assume roles not directly connected with the settlement. Their readiness to do so may be explained by their ambition to play a central role in the building of the country in collaboration with the Histadruth and by the desire to consolidate the kibbutz. They realized that it would be impossible to grow and develop as a significant social force within the framework of one settlement, and that this one settlement, limited as it was in its area, in its financial resources, and in the agricultural know-how of its members, could never hope to grow at the rate necessary to absorb a large number of new immigrants.

The solution appeared to lie in the establishment of similar kibbutzim, among whom Ein Harod would maintain a central position, further strengthened by activities outside the settlement, contributing to political strength and at the same time performing vital national functions. This expansionist orientation reflected the twofold desire of becoming the kernel of a strong organization, and of greater social and national participation. The exponents of this approach sensed the urgent social need for numerous new frameworks of immigrant absorption that would not require the immediate allocation of funds for settlement, but would at the same time be a potential reservoir for such settlement in the future. They also recognized the importance of placing at the disposal of the Histadruth a loyal instrument in its labor struggles to replace the Battalion.

The minority view held that at this stage it was essential to direct all the available resources and energies towards the consolidation and further development of the settlement itself; by directing part of them to other functions the very success of the village was endangered. Having learnt a lesson from the experience with the Battalion, this group feared that the separation of a work company from the settlement would hinder the social integration of all the members of the kibbutz and substitute a partnership based solely on political affiliation for a full partnership in all aspects of life.

When the conflict between the two opposing sections, which cut across group lines, grew more bitter, the minority—on the basis of the precedent—demanded a renewed partition of the settlement. This time, however, the Histadruth was not prepared to endorse their claim. The idea of the large kibbutz had already gained acceptance, and the Histadruth was anxious to grant its support to the group that had accepted its challenge and expressed its willingness to cooperate. This conflict was finally resolved by a decision that the minority leave the kibbutz to join one of the existing moshavim, or cooperative settlements.

The departure of these members in 1926 left Ein Harod united in the orientation towards social participation, and the issue of such participation was now solved once and for all. It was, however, not the solution of the Battalion, adopted after the Battalion itself had left: for while in Gdud Haavoda social participation meant the subordination of Ein Harod to an external political center, here the program provided for the settlement to be itself the center of initiative. This resolution served as the basis of the Hakibbutz Hameuchad movement and of the creation of a number of work companies—the plugot of Ein Harod—scattered all over the country. The organization of these companies was on the lines of the Battalion; it carried out similar tasks and was characterized by a similar degree of mobility and flexibility and a pioneering orientation. Here, however, the ultimate purpose was the transformation of the companies, when the time came, into new settlements. This part of the story, though, is taken up more fully in Chapter 6.

It can thus be seen that the first five years of Ein Harod were years of internal struggle, during which the settlement had little chance of crystallizing socially. A similar situation existed also in the economic sphere, which was affected not only by the division of resources and by the social crises as such, but also by the lack of support by the settlement institutions.

The relevant documents reveal the settlers' continuous complaints that Ein Harod was being financially discriminated against in comparison to other settlements, both collective and cooperative, which were established at the same time. The funds supplied in these first years by the national institutions were not sufficient to meet the needs of the settlement, although they were granted under convenient terms of repayment. Moreover, the funds were not made available in a lump sum but in a series of small allotments. This practice was not adopted for the moshavim, which then received the necessary sums during the first stages of settlement and thus succeeded in consolidating in a relatively short period. This factor (and others which will be discussed later) caused a delay in the development of the settlement, restricted its productive capacity, and prevented the rational utilization and investment of the available funds, which had to be used largely to cover current expenses.[12]

Thus, we find that in 1926, Ein Harod and Tel Yoseph, both large kvutzot, received much less capital funds than other, smaller collectives established in the same area and at the same time: Ein Harod and Tel Yoseph received in 1926 £P 231 and £P 264 respectively, when Hephziba received £P 499, Bet Alpha and Geva received £P 433 and £P 645 respectively.

The crises which affected the settlement in its first years were interpreted by the Zionist institutions as proof of the settlers' instability and of the impracticability of the pattern as such, in which they had not believed in the first place. Dependence on outside authorities, whose unsympathetic attitude continued till the thirties, and the bitterness arising from the feelings of discrimination are clearly reflected in the following letter sent in 1930 by a number of leading members of Ein Harod to the Executive of the Zionist Organization:

> Are there, indeed, no limits to the victimization of the public and the settlement, and will its fate be entrusted forever to the hands of a capricious administration? The Histadruth should compel the Executive of the Zionist Organization to remove the obstacles hampering the development of our settlement. For nine years now our settlement has been struggling for survival. Internal and external worries have left their mark. Others, who settled on the land at the same time as we did, have long since attained a reasonable degree of consolidation. No other settlement has been obliged to face the pressures that we were obliged to face. But we have passed the test and succeeded in overcoming severe crises. Now we are on the verge of social and economic development, and we cannot tolerate that the public, whose emissaries and pioneers in settlement we conceive ourselves to be, should be indifferent to us and permit us to exhaust our strength in vain.

Another factor that affected the development of the settlement during its initial period was the problem of land. Early in 1921 an area of 5,750 acres was purchased by the Jewish National Fund in the eastern part of the Jezreel Valley. The founders of Ein Harod demanded the entire area for the realization of the ideal of the large kvutza, which was to be composed of three sites. The Histadruth, even though it endorsed the experiment at its Agricultural Conference of 1921, could not comply with the settlers' demands for the allocation of the entire area, since pressure for land was great, and other groups who had been waiting for long periods were entitled to their share. After lengthy debates between the Histadruth and the Battalion, the area was divided: 3,250 acres were allotted to the large kvutza and 2,500 acres to four other groups.[13]

After the division of the settlement, the minority at Ein Harod was left with an area of 1,500 acres, too small in proportion to the number of settlers, and the problem was aggravated by the distance and dispersal of the land, which caused difficulties and unnecessary expenses. And since land, like capital, was scarce, and the priority of Ein Harod considered low, the situation did not improve till the beginning of the second decade.

The Third Period—Consolidation and Growth (1926–1948)

As was seen above, the initial five years of Ein Harod was a period of social strife and crisis, and 1926 is thus perhaps the first year of stability. It is from this turning point, therefore, that our proper story of institution and society building in Ein Harod begins, even though the grounds for the institutional structure were actually laid earlier, and even though the economic take-off had to wait five more years.

Like all settlements established by the Zionist movement, Ein Harod was built on land owned by the national institutions and thus protected against alienation, speculation, and misuse. These restrictions, however, even though they had few legal teeth, were actually irrelevant to the kibbutz; its forty-nine year tenure—renewable and subject to symbolic payment only—presented complete security and freedom of use and had all the advantages, and few of the liabilities, of the freehold. The same applied also to financing, which was institutionally public, assuring initial investment as well as regular financial care and responsibility, but at the same time leaving the freedom to utilize regular commercial sources.

The pattern of mixed farming was introduced into Ein Harod in accordance with the ideological blueprint, and its adoption fulfilled expectations. It assured stability in the allocation of work and was well adapted to the realization of the principle of self-labor, as it enabled, more than any other agricultural pattern, the equal distribution of work hours over the entire year. It guaranteed a considerable degree of economic stability by reducing dependence on the success of any single agricultural branch, and provided for the utilization of a considerable percentage of the produce—up to 40 per cent—for home consumption. This meant that a stable internal market existed for the second-and third-quality produce that might have caused losses in the open market. It also reduced marketing expenses to a considerable extent, as a large proportion of the produce was utilized in the settlement itself.

The initial branches developed at Ein Harod were those usually associated with mixed farming: field and fodder crops, orchards, vegetable garden, chicken runs, and dairy. However, new and unconventional branches were soon introduced.[14] The settlement indeed served as a testing ground for experimental crops new to the country in general and to the Jezreel Valley in particular. An example of this were the citrus groves planted in the heavy soil of the valley in disregard of expert opinion. The experiment was successful and paved the way for the development of the citrus branch in the area. Other agricultural innovations were fodder under irrigation, plant nurseries, and beehives.

In addition to the mixed farm—itself constantly adjusted, changed, and improved—Ein Harod put into practice, from the very beginning, the concept of an integrated economy. The first workshops were established prior to the split with the Battalion, and they were designed for sewing, production of shoes, and carpentry. Somewhat later, a blacksmith and fitting shops were added. The original aim of these workshops was to supply the settlement's need for worktools and various appliances. However, shortly after the union with Chavurat Haemek, the workshops began to produce for the external market, although as yet on a small scale.[15]

Ein Harod was also the first settlement to mechanize its agriculture. A short time after its establishment a tractor was bought, and within a few years all field crops were harvested by machinery. An incubator for the poultry runs, another laborsaving device and an innovation unknown to other settlements at that period, was purchased during the first period.

The workshops were even more remarkable for their efforts in rationalization and mechanization than were the agricultural branches. As they were experimental and quite new in the framework of an agricultural settlement, constant attempts were made to render them economically profitable and market oriented. The workshops expanded and developed and became an important factor in the kibbutz's economy.[16]

As we saw, much of this drive to constant innovation was embedded in the basic ideology of the settlers which, conceiving of rationality, growth, and achievement as values in themselves, fostered improvement and experimentation. This original orientation, however, was reinforced by the challenge of the stubborn disbelief that the large kibbutz faced among the settlement authorities. This disbelief extended, as was mentioned, to the potential of the pattern as such, but also to the qualifications of the group to establish and maintain a viable community. Zionist authorities and agricultural experts alike thus exerted considerable pressure on the settlement to limit membership, even to the point of threatening to cut off funds; doubts were similarly expressed about the diversification of the economy into new farm and nonfarm branches. In order to continue the experiment, the kibbutz was obliged to prove its ability to support a large population, and no effort was spared to take advantage of any possible opening.

The main condition for successful growth and development, however, seemed to lie in the agrarian structure itself; the collective unit of production, embodying a large and centrally managed reservoir of skilled manpower, was an ideal institutional arrangement for maximum mobilization and rational utilization of resources, and for the quickest diffusion of innovation. Full advantage was taken of this potential for the efficient planning of various branches through constant modification in response to increased experience and conditions. Conversely, skilled workers could be allocated and relocated according to demands. The collective unit of production also facilitated, of course, the new experiments as well as the use of mechanization; and it compensated thus to a large extent for the initial scarcity of land. Of interest in this respect was the fact that in Ein Harod the sphere of services to the members (usually the least dynamic as well as the least desirable) did not lag behind the productive one. In the large-scale services required by the numerous population, any improvement meant a considerable saving in investment and man-hours, which could profitably be directed to other branches. The kibbutz was thus impelled to apply the economics of size also to the sphere of consumption. This may be illustrated by the washing machine introduced in the Ein Harod laundry in the twenties, which was the first, if not in the country, then certainly among the farm population.

During the first decade of its existence, Ein Harod faced several ecological problems. The settlement was still located on a temporary site, and while the fields were already close to the permanent location, for financial reasons the farm buildings and living quarters could not be transferred for a long time. This compelled settlers to cover long distances daily in order to get to the fields and plantations, and caused economic loss and considerable inconvenience to the members, some of whom even had to stay outside the settlement, near the fields, for the larger part of the week. The ten-year period of temporary living quarters at Ein Harod was extremely lengthy in comparison to such periods in other settlements that were established at the same time.[17] The temporariness of the arrangements also hindered the introduction of improvements in the living quarters and in the farm buildings, because all expenditure of funds on temporary projects was—rightly—considered wasteful. The personal and economic problems were completely solved only after the transfer of the entire settlement to its permanent site.

By contrast, the principles of the village lay-out, incorporating the "gathered" ecological pattern, already were established during this period; as can be seen from the photograph and map, they were essentially maintained on the new site. The dairy, poultry, workshop, and other branches and services were thus situated close to the residential quarters and services, within the kibbutz compound, while the field crops and plantations were radiated out from the center of the settlement. The problems of transportation, which were quite severe during the first years and still existed with the later increase in the cultivated area, were solved with the introduction of mechanization, especially the use of tractors to transport people to work.

Another ecological arrangement, deriving from the principle of collectivity in the provision of services, was the separation of the living quarters of the members and the central services such as the communal dining hall, the clothes stores, the laundry, etc. The shower rooms and conveniences were also separated from the rooms for practical reasons, because this arrangement was less expensive than building them adjacent to the rooms. This required the establishment of all these services in the center as close as possible to the rooms, in order to reduce the amount of daily walking. This problem could not be tackled as long as the settlement was on the temporary site. However, after the transfer to the permanent site it was solved to the satisfaction of the members. Although a member of the kibbutz still had to put up with certain inconveniences, because he had to walk considerable distances every day in order to use the services that usually are provided at home, this was accepted without any hesitation as being in line with the communal values. For the same reasons, members were prepared to accept the lack of privacy involved in the use of communal shower rooms.

All these disadvantages in services, however, seem to have been balanced by the economically vital fact that the collective

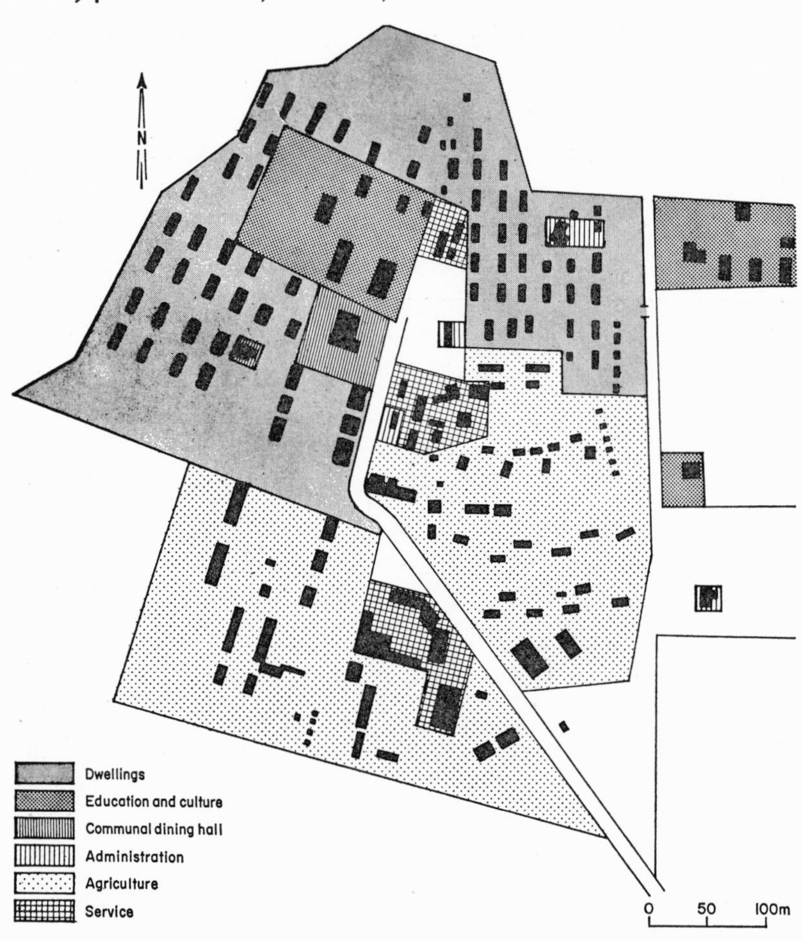

Map 8. Ein Harod, 1943

unit of cultivation eliminated the main drawback of the gathered village, which we shall observe in Nahalal: the inevitability of dispersion of the land units. The kibbutz had most of the advantages and few of the drawbacks of the gathered form of settlement. The character and development of the agrarian structure of Ein Harod, particularly its land-use system, are illustrated in Plate II and Map 8, of 1946 and 1943 respectively.

As was mentioned above, the year 1931 constituted an economic breakthrough for Ein Harod. The obvious success of the settlement, together with the support it received from the Histadruth,[18] combined to change the attitude of the settlement authorities to Ein Harod; in 1931 the kibbutz not only received additional land but was finally moved to the permanent site. The Settlement Department, under the influence of Dr. Ruppin, also accepted Ein Harod's financial claims and allocated the funds necessary for consolidation. In 1933 the settlement authorities erased the largest portion of the settlement's debts to them. The members defined this agreement between the kibbutz and the Settlement Department as the "point when Ein Harod no longer

Table 11. Capital accumulated in Ein Harod from agricultural profits, 1934–1939 (in £P)

Year	£P
1934	18
1935	1810
1936	1308
1937	923
1938	402
1939	2793

Compiled from *Conditions at Ein Harod and Developmental Trends*, in Hebrew (Kibbutz Ein Harod), p. 91.

faced the dangers of collapse that had threatened it during the first ten years of its existence." [19]

Financial dependence on the settlement authorities was gradually reduced during the new period until it completely disappeared, although loans were still required from various sources such as banks and the Histadruth institutions. This system of financing the development of an agricultural settlement by way of loans to be repaid by further loans was to become a regular feature of the collective settlements in Palestine, including Ein Harod. However, there is an essential difference between this period and the preceding one, for now the settlement became financially independent and paid its own way.

Table 11 indicates the over-all progress towards economic

independence made in Ein Harod after the supply of basic production factors.

In 1938 the settlement had 4,060 acres of land at its disposal, 3,840 of them cultivated and 440 under irrigation. The area of

Table 12. Gross income in Ein Harod, 1932 and 1938

	Gross income			
	In 1932		In 1938	
Branches	£P	%	£P	%
Field crops	2,619	13.0	5,078	10.1
Fodder crops	1,823	9.1	3,047	6.1
Vegetables	624	3.1	2,109	4.2
Plantations	1,695	8.4		
Gardens	157	0.8	556	1.1
Nurseries	1,528	7.6	1,377	2.8
Dairy	5,349	26.6	9,842	19.7
Goats			443	0.9
Poultry	1,814	9.0	2,885	5.8
Honey	171	0.9	316	0.6
Orchards			4,027	8.0
Vineyard			3,035	6.1
Fruit trees			844	1.7
Olives			411	0.8
Total, all branches	15,780	78.5	33,970	67.9
Workshops	1,993	10.3	10,418	20.8
Paid labor	260	9.9	3,968	7.9
Others	2,083	1.3	80	0.2
Total	4,336	21.5	14,463	28.9
Total in settlement	20,116	100.0	48,436	96.8
Additional outside investments	–	–	1,600	3.2
Grand total	20,116	100.0	50,036	100.0

Compiled from *Conditions at Ein Harod and Developmental Trends*, p. 88.

land per worker was now sixteen acres; however, five of those were located at a considerable distance from the settlement in areas that had been purchased in the Zevullun Valley.

Table 12 shows the income obtained from the various branches in the years 1932 and 1938. (See also Chart 1.)

The branches producing most income are the dairy, the field and fodder crops, the poultry runs, and the plantations. In 1932 the income from the dairy made up 26 per cent of the total, field crops 13 per cent, fodder 9 per cent, and plantations 8 per cent. Income from other branches was lower. In 1938 we do not find any variation in the structure of the branches and no change in

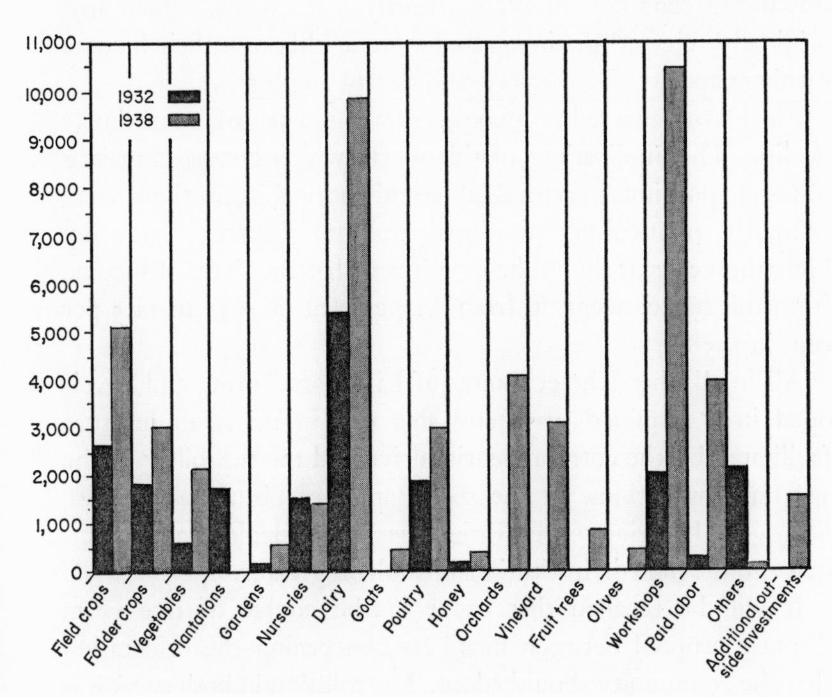

Chart 1. Gross income in Ein Harod in 1932 and 1938 (in £P)

the branch that contributes most to the settlement's economy. However, there is a change in the percentage of the income that each branch contributes; and now the dairy supplies 19 per cent, field crops 10 per cent, orchards 8 per cent, vineyards and fodder 6 per cent, and the poultry runs 5.5 per cent of the total.

Thus, we can see that the mixed-farming structure remained stable and the relative weight of each branch remained constant. Progress was made in balancing the various branches, so that those producing considerable income were increased, while the dominance of any one specific branch decreased. The increase in income percentage from the workshops during this six-year period should also be mentioned. This doubled, which illustrates the development of these branches and the success in the implementation of the principle of an integrated economy.

It is interesting to note that by 1938 workshops also penetrated the economy of other, nearby settlements, which had originally rejected the integrated system, although their income earning capacity did not yet reach that of Ein Harod.

Ein Harod also led in income from the work of its members outside. The implications of this work, which consisted mainly of public positions in central labor and national institutions, were primarily political-organizational and are discussed later on. Here, however, it should be mentioned that the share of income from this source went up from 8.1 per cent in 1932 to 12.5 per cent in 1958.

All in all, then, the economy of Ein Harod grew and developed in a sustained way, and this was in no small measure facilitated by the entrepreneurial drive and the flexibility of the agrarian institutions. In the social and organizational spheres, however, the drive to innovation was more ideologically inhibited and the built-in institutional problems greater.

It will be recalled that twice in the course of five years disputes erupted between members concerning the orientation that the community should adopt. The solution in both cases was similar: the protagonists could no longer live together, and one of the groups was compelled to leave. This resulted in the first instance in the division of the village, and in the second in the departure of the minority to a moshav.

This phenomenon, while unusual in the set-up of a farm settlement, resulted from characteristics built into the kibbutz pattern. First, there was the extreme preoccupation of the kibbutz members with ideology; issues were not only defined in ideological terms, but also with great rigidity, admitting of no compromise or neutrality. Ideological commitment, moreover, extended to broad areas, so the settler was articulate not only on community problems, but on general social and political questions as well. The majority indeed regarded the kibbutz itself not just from the point of view of individual and group interests, but as a focus and a reflection of the needs and aspirations of the Yishuv. The broad ideological spectrum, of course, enormously increased the potential areas of conflict, and this was further aggravated by the high degree of consensus required by the community. Thus, while Ein Harod and similar settlements did not uphold the value of total ideological collectivism or uniformity adopted by Hashomer Hatzair movement, its scope for differences of opinion was nevertheless small and did not extend to issues considered vital. The members of Ein Harod, however, due not only to historical necessity but also to the kibbutz's open-door policy, seldom constituted a homogeneous and a preselected group, conforming to one view. Also, the collective structure of the kibbutz, with its intensive interaction and participation, and its total social and organizational system containing within one framework all institutional spheres, prevented the segregation and the separation of groups and issues and the isolation of potential foci of conflict.

Initially, this situation even fit into the general conditions of colonization, which then made possible—if not encouraged—the leaving of individual members and groups. The major part of the country was as yet unsettled, and in spite of the difficulties of acquiring land, vast empty spaces were waiting to be occupied. This was also a time of social and colonizing experimentation, before patterns were fixed, sanctified, and entrenched; the scope for innovation—in ideology or practice—was great. Essentially, however, social cleavage on ideological basis is inherent in the large kibbutz. While the crisis of division occurred in Ein Harod again only after 1948,[20] the problem has been endemic, with members unable to conform leaving or being asked to leave.

The most basic problem of Ein Harod, however, was the upholding of its own collective and communal values, the maintenance of social integration, and the creation of proper motivation among its members, in the light of its dynamic and developing economy, its growing society, and its open-door recruitment policy. That it was essentially able to limit dissatisfaction and desertion, and protect group solidarity, while at the same time upholding collective and egalitarian ideals, is perhaps the most striking feature of the society examined. This was achieved, it seems, not only through intensive socialization and cultural activity, but chiefly by a sensitive touch on the pulse of life, a constant—though controlled—reinterpretation of principles, and secondary institutionalization.

These mechanisms stand out most clearly in five interrelated problem areas, reflecting the process of social change, experimentation, adjustment, and innovation: transition from face-to-face, *Bund*-like,[21] and small community life, to a large crystallization of solidary subsystems, and the strengthening of the place of the family within the collective; role-differentiation and specialization in the economic sphere; manpower recruitment; a dramatically rising standard of living; and transition from direct democracy to an articulated authority structure.

The feature that stands out most in Ein Harod's social history was its demographic process, characterized by growth and diversity. Ein Harod remained loyal to the principle of a large and a growing membership, and following the improvement in the economic situation, the settlement absorbed immigrant and youth groups. By the fifteenth year of its existence, there were 404 members at Ein Harod; in all, the population numbered 742, including the children. This population continued to grow, by degrees, and by 1940 there were in the village 929 people, including 450 members.[22]

Table 13 indicates the population composition of Ein Harod, in comparison to other kibbutzim of the area, all of which were founded at the same period.

The figures offer a comparison between the populations of the large kvutza (Ein Harod and Tel Yoseph), a typical small kvutza (Geva), and two medium-sized kibbutzim (Hephziba and Bet Alpha, later incorporated into Hakibbutz Haartzi, the settlement organization of the more radical Hashomer Hatzair

Table 13. Population at the close of 1938 in five kibbutzim (in absolute figures and per cent)

	Ein Harod		Tel Yoseph		Hephziba		Bet Alpha		Geva	
Category	No.	%	No.	%	No.	%	No.	%	No.	%
Members and candidates	417	52.4	309	52.5	83	48.2	137	52.3	100	40.3
Temporary and training groups	19	2.4	21	3.6	22	12.9	20	7.6	29	11.6
Professional employees	4	0.5	13	2.2	2	1.2	5	1.9	4	1.6
Locally born	246	30.9	171	29.1	58	33.7	88	33.6	68	27.3
Born outside	6	0.8	7	1.2	–	–	–	–	11	4.4
Dependents	44	5.5	28	4.8	7	4.0	12	4.6	17	6.8
Migrant youth	60	7.5	39	6.6	–	–	–	–	20	8.0
Total	796	100.0	588	100.0	172	100.0	262	100.0	249	100.0

Compiled from *Conditions at Ein Harod and Developmental Trends*, p. 78.

movement). As can be seen, the population of Ein Harod was considerably larger than that of any other settlement, as was the percentage of children and dependents. This process of demographic growth also brought about increasing social heterogeneity, which is clearly reflected in Table 14.

The heterogeneity of the population extended, however, not only to country of origin, but also to basic differences in background and preparation for settlement. A decisive step was

Table 14. Ein Harod membership by countries of origin (in per cent)

Country of origin	Per cent
Poland	30.9
Russia	28.6
Germany	26.7
Lithuania	3.4
Israel	2.5
Roumania	3.2
Austria	1.3
U.S.A.	0.5
Czechoslovakia	0.2
Others	2.7

Compiled from S. Landshuth, *The Kvutza: a Sociological Analysis of Kibbutz Settlement*, in Hebrew (The Little Zionist Library, 1944), p. 66.

taken in this direction in 1934, when the first Youth Aliya began arriving from Germany. The absorption of such groups was a striking innovation, since it meant accepting school-age children, the majority of whom came, moreover, from middle-class, assimilated homes, with little pioneering and nationalistic values. The kibbutz had not only to socialize the youth into its own ideology and to provide intensive Hebrew and national education, but also to serve as a substitute for the lost family and home. The fulfilment of this national function, in which Ein Harod stood alone for a long time, was successful; and its "graduates" were the first to establish an immigrant youth settlement in the country. Ein Harod continued to absorb immigrant youth groups, and even before the close of World War II, it was the foremost settlement in taking in survivors of the Nazi holocaust. Since the war, other groups of youth from Europe, as well as the East, have been absorbed.[23] This later phase was even more complicated than the early one, since here the children came from less developed countries and from more traditional backgrounds.[24] While Ein Harod was willing to make the necessary changes in its program, socialization to the kibbutz values was less successful, and the majority eventually left.

Social heterogeneity in Ein Harod was thus a function of the policy to create a large and a varied community, and of the acceptance of immigrant-absorption tasks. The open-door policy, however, compounded by processes of role-differentiation, eliminated the intensive face-to-face interaction between all members that had been the basis of community integration in the small kvutza and in Ein Harod itself in the early phase. Interaction in Ein Harod thus became based on each member's involvement in a wide network of contacts arising out of his social characteristics and various roles: as neighbor, friend, age-mate, fellow countryman, participant in a work group, a member of a committee, and others. A considerable time was spent every day in informal groupings, notably during meals in the communal dining hall. Later, additional centers for group activity began to appear: rooms reserved for committee meetings, a center for culture and entertainment, the art museum, and others.

However, the crucial quality of this pluralistic structure, embodying as it did temporary and permanent groupings and partial aggregates within the total situation, was that it did not interfere with the creation of an over-all solidarity. Indeed, the different constellations interacted with each other directly or through interlinking and overlappings and very often formed and acted in response to the collective's larger objectives, so the kibbutz's main social feature and advantage, its mobilization potential, was not apparently impaired. On the contrary, the variety of new elements with different backgrounds and of different ages contributed to a flow of potential innovators and

Table 15. Seniority of membership, 1938

	Ein Harod		Tel Yoseph		Geva		Hephziba		Bet Alpha	
Time	Members	%	Members	%	Members	%	Members	%	Members	%
Less than 2 yrs.	75	18	57	3	19	20	1	1	8	6
2–5 yrs.	85	20	72	28	28	30	23	31	32	23
5–10 yrs.	23	6	73	28	24	25	26	34	17	13
More than 10 yrs.	235	56	107	41	25	25	26	34	80	58
Total	418	100	309	100	96	100	76	100	137	100

Source: Conditions at Ein Harod and Developmental Trends, p. 75.

constituted, as we shall see later, a reservoir for a regular circulation of élites. This factor manifests itself also in the relative stability of the population, belying the general skepticism as to the social viability of the large kvutza, which was given by the settlement authorities the name of a "temporary arrangement for the introduction of pioneers to settlement work."

Table 15 shows the high percentage in Ein Harod of veteran members and indicates the low rate of turnover.

A major change occurred similarly in the relationship between the collective and the family. It was, of course, a primary tenet of communal life that the collective took over from the family the functions of socialization and placement, directing the education and the role-allocation of the young generation. This fundamental principle was maintained in Ein Harod and reflected in a central nursery (where mothers breast fed their babies) and daylong schooling and care. Children spent most of their time in a special children's compound, under the supervision of nurses, kindergarten teachers, and house mothers, and were formed into organized age groups. However, while in most collectives (including the small kvutza and the kibbutzim of Hashomer Hatzair) the children also slept in this compound and were with their family only during special visiting hours, Ein Harod early adopted the policy of allowing children to sleep in their parents' quarters. They were thus with the parents from afternoon till morning, and the family framework was largely preserved, eliminating tensions destined to arise in other settlements.[25] Indeed, building the family houses was in Ein Harod one of the early goals, and it preceded many other improvements in private and public consumption made possible by economic success.

Ein Harod seems, as a matter of fact, to have alleviated the tensions between the demands of the collective and the individual's desire for a private domain of his own, where he could spend time with his immediate family or friends, or pursue a hobby of his own. Obviously, the solution to this conflict was achieved differently by different individuals; while one showed a tendency to retire more into the private sphere, another was more willing to assume social responsibilities and to take part in collective activities. The crucial fact, however, was that a wide margin for choice and a scope for different preferences existed.

At the same time, the general climate of Ein Harod, in comparison to many other collective settlements, was one of high social participation and interest of the general membership in the affairs of the settlement and of the movement; this is one of the main reasons why Ein Harod managed to fulfil its ideology and to maintain its élite role. The usual schedule of a member of the settlement was thus to devote the morning and early afternoon, with a break at noon, to his work; the afternoon hours were mostly spent in the family circle and devoted mainly to talking and playing with the children, who had come home to their parents' room; the evenings were for private visiting with friends, and other individual activities. However, several evenings a week belonged to the public domain and to attending committee meetings, participating in cultural activities of the entire kibbutz, or taking part in a general assembly, a political discussion, etc.

Considerable flexibility also existed in the area of division of labor. As we saw, the very concept of the integrated and diversified economy, with its emphasis on development and rationality, laid the basis for a differentiated and specialized role structure. The drive to entrepreneurship was further reinforced in this respect by the demographic growth and by social heterogeneity.

In the early days of the settlement, the majority of the settlers were employed in unspecialized jobs, such as clearing and construction; however, after a brief period, a certain differentiation occurred in the economic sphere. Various agricultural and other areas were given the name of branches, and members began to specialize in one particular branch, such as the dairy, the orchards, field crops, or the carpentry shop. Every such branch was managed by a member with some experience and know-how (usually from the veteran group) who gathered a core of permanent workers around him.

Gradually, during the next few years, an increasing number of members transferred from unspecialized jobs to specialized ones. This created, first of all, a major dilemma in role-allocation, a potential conflict between the personal preferences of a member to work in a particular branch and the requirements of the kibbutz economy, which demanded a given number of workers in the various branches, and which often necessitated the transfer of a person of ability to a particular branch in order to raise its standard. The kibbutz ideology, it will be remembered, stressed the importance of employing the members in areas of work that corresponded to their wishes and talents, so as to allow for maximum individual fulfilment; naturally the members themselves exercised pressure to be placed in their preferred branches. However, this could not be regulated solely according to personal wishes and predilections, as such a situation could easily result in the neglect of less popular branches that were nevertheless essential for the kibbutz economy. Most unpopular were the services, particularly the nonspecialized tasks of cleaning, serving the meals, dishwashing, helping in the kitchen, etc. These jobs were obviously regarded as unsatisfying to the individual member, as well as of low prestige; the obvious solution of most family-based units of production and consumption—allocating the manual and the domestic jobs to the women—was not at first acceptable in an egalitarian society.[26]

A partial solution to this problem was job rotation, which not only assured the performance of all necessary tasks, but also symbolized the value and the practice of equality, for the most specialized and respected members did their share of the least specialized and respected tasks. The mechanism devised to deal with this problem and to provide workers for peak agricultural season by mobilization from quieter branches was a roster. This roster was drawn up by a special committee that was the supreme authority in all matters concerning job allocation and acted in the name of the general assembly. The committee thus could compel a member to work in any branch required for a maximum period of two years, after which he was entitled to return to his specialized role.

In this connection, the norm of an eight-hour work day was established, as was the norm of one rest day per week. This rest day was not necessarily the *Shabbat*, the Jewish day of rest. Every member chose one rest day per week, taking into account the requirements of his specific branch and his personal preference. As for the eight-hour work day, this was not always maintained in practice, because of the general shortage of manpower and because at the peak seasons certain branches required more than a full day of work from the members employed in them. However, the kibbutz aspired to the norm, and with the increase in mechanization and the rationalization of work methods, it was approached at most times.

Another interesting mechanism in the sphere of division of labor was a frequent distinction between the formal head of a branch and its real expert. The first was thus often rotated, preventing the crystallization of an explicit managerial élite, while the other provided continuity of know-how.

All these arrangements, some institutionalized and some not, combined to narrow the inherent contradiction between differentiation and egalitarianism. They did not, it is true, prevent the

Table 16. Allocation of jobs by type in Ein Harod, 1940

Branch	Workers	Per cent
Agricultural	158	31.2
Various agricultural and timber-cutting	11	2.2
Workshops	55	10.9
Outside jobs, construction	31	6.1
Watchmen, general duties	23	4.5
(Total working)	(278)	(54.9)
Services	169	33.4
Leave, illness, childbirth	59	11.7
Total	506	100.0

Source: Conditions at Ein Harod and Developmental Trends, p. 12.

emergence of social stratification within the population. However, this stratification was community oriented and based on public activity and economic and organizational abilities; it was neither permanent nor formally rewarded, and it did not create any distance between roles and groups. This feature, together with the high sense of service and the general ideological commitment, thus accommodated somewhat different aspirations and responsibilities, together with nondifferential rewards, and provided motivation for the performance of many diverse roles.[27]

Table 16 gives an idea of the division of work among members of the settlement.

Intimately related to the problem of division of labor was the question of manpower in general, in the light of considerable outside activity on the one hand, and prohibition of hired labor on the other. As we saw, wide social participation was one of the primary aims adopted by Ein Harod in 1926, which meant the assumption of various pioneering, political, and security tasks in national and labor institutions. Mobilization for these tasks was also centrally organized, the movement deciding on the share of every kibbutz, and each kibbutz allocating specific members.

This export of capable members of the kibbutz to perform élite roles in the larger society fulfilled an important function: it served as an outlet for talent and ambition that could not find proper scope in the kibbutz itself. This mechanism also provided another solution to the tension inherent in the egalitarian pattern of kibbutz life; for although the members were all entitled to the same material rewards for their work, regardless of its particular nature, the performance of élite roles outside the kibbutz served as a significant compensation for the lack of material rewards in the kibbutz society. The member working outside the kibbutz not only performed interesting work that gave him satisfaction, prestige, and sometimes power, but also lived a freer life, broadened his horizons, and was entitled to spend a part of the salary that he received for his work according to his personal preferences. Subject to qualification, these jobs were rotated too. In time, though, a class of professional functionaries seems to have arisen, whose integration with the kibbutz was attenuated and who were likely to become a focus of jealousy and of feelings of relative deprivation.

External functions also served, however, to increase manpower shortage in the kibbutz, a shortage which could only partly be offset by the advantages of the collective unit of production and by central mobilization and allocation. The principle of self-labor, specified in the ideological blueprint, was largely maintained by the kibbutz, at least initially, although this caused considerable difficulties for the policy of expansion and development. The sociological implication of this principle was that the kibbutz could not maintain a mutual system of manpower exchange with the larger society. For its part, it was prepared to allocate manpower to perform various roles in the larger society, but it would agree to accept manpower from the outside only on condition that it was recruited through the ideological channels and was committed, at least temporarily, to the collective ideal. By this means, the kibbutz managed to preserve its wide social participation and its ideological uniqueness.

However, it had to solve its problems of manpower shortage, and for this purpose it evolved various social set-ups through which it obtained the necessary manpower on a temporary basis. The framework here was the youth movement committed to the kibbutz ideology, whose members, from the Diaspora and from the country, organized groups for the purpose of establishing agricultural settlements. Since they were not yet prepared for settlement, lacking any practical training and know-how in agriculture, and since funds and land were often not available, the groups spent a period averaging two years at Ein Harod and other kibbutzim, where they worked and received their agricultural training.

All in all, though, lack of sufficient manpower constituted the most important check on the entrepreneurial drive of Ein Harod; and when the various solutions proved insufficient, the settlement resorted to hired labor. This, however, was always considered ideologically deviant and never received sanction.[28]

The problem of consumption created considerable tension, brought to the fore by economic growth. As we saw above, a rising standard of living was encouraged at Ein Harod, which never preached austerity for its own sake. The growing prosperity enabled the settlement to increase substantially the standard of services and consumption on both individual and community level, notably in family and public buildings, personal and recreation budgets, and similar spheres.

Economic consolidation thus facilitated the building of spacious family houses, made of stone and with conveniences attached, which were allocated according to a priority list based on length of residence in the kibbutz and family status. If, as a result of changing conditions and growing demands from the rank and file, innovations had to be introduced, this always involved a prolonged struggle, in which the different views were aired orally and in writing and grounded in ideological arguments. Communal egalitarianism circumscribed consumption, and most demands sought greater individual freedom of decision and greater differentiation in consumption patterns.

Issues of this type were particularly prominent in the area of private consumption, where pressures were exerted by the individual members—particularly by the women. Here the élite of the kibbutz and of the movement alike guarded against the encroachment on the principles of egalitarianism by the growing desire for a more individualistic pattern of life and greater material rewards. A hotly debated issue was a personal budget for every member for articles of clothing and everyday necessities, versus the allottments to every member of a fixed amount of articles from the central supply store of the kibbutz. This, however, is only an illustration; issues of this type constantly arose in every area of consumption and constituted a constant irritation. The solutions reached usually permitted a minimum gratification of individual desires, while at the same time strictly preserving the egalitarian pattern. This preservation was considered so important that the resolutions were usually discussed and adopted by the movement as a whole.

Finally, cutting across all the problems mentioned, and focusing them, there was the articulation of the authority structure. The Battalion initiated the original patterns of organization at the kibbutz, since its members constituted the majority of the settlers during the first phase and brought with them the organizational patterns evolved in the course of their activities in the country. As already mentioned, the Battalion was organized in work companies, and these companies were formed in life cells, comprising ten members in each cell. The cells were based on ties of mutual friendship and common background between members, and their sphere of activity was unlimited. One representative of each cell was elected to represent his companions in a central body, which was called the company council and was entrusted with the management of all the affairs of the work company. This organizational pattern combined military regimentation and direct democracy, and as such was perfectly adapted to the requirements of the Battalion, a semimilitary organization with a low degree of differentiation both in the jobs and in the requirements of its members.

This structure was retained during the first period of settlement before specialization and differentiation became an important factor in kibbutz society. However, a new and supreme institution was introduced into the social system, the general assembly. The main purpose of this institution was to achieve maximum integration of all the members and to communicate to them all developments in various spheres of activity, which proliferated as life in the settlement grew more dynamic and varied. At that time, there was still no differentiation between the economic and social spheres, and the general assembly dealt equally with both. The council, inherited from the Battalion, was not discarded but retained its function of managing the affairs of the settlement; however, the division of authority between the two bodies was not clearly defined, and decisions were made in both of them simultaneously.

As a result of the split with the Battalion, certain changes occurred. The life cells disappeared, since they no longer answered the requirements of a differentiated society. With their disappearance, the council ceased to exist, and the general assembly became supreme and assumed the functions of policymaking, communication, and social control.

The first signs of organizational specialization occurred with regard to certain executive roles, mainly economic, such as treasurer and purchasing officer. These roles were not as yet institutionalized and were totally identified with the individual performing them, who assumed the job by right of special talents and competence. In this way, the principle of direct democracy was still maintained, and all important decisions were made by the general assembly itself. However, with the progress of differentiation, discussions became less general, more technical, and confined to certain areas, a factor which contributed to the decline in the number of members participating.[29] Parallel to this, other departures from direct democracy were institutionalized: the creation of specific roles based on a high degree of specialization and competence; and the formalization of various functions previously performed on a voluntary basis.

The kibbutz strove to fill these positions according to the principles of democratic election and rotation. A minority of them were considered full-time occupations, and the members performing them were not expected to work at any other jobs during their period of office. The majority, however, were performed by members after work. New governing institutions, smaller in number and more centralized than the general assembly, were evolved to manage the affairs of the kibbutz—and they exist to this day.[30] The most important organ in charge of the daily affairs of the settlement was the active secretariat, comprising eight members. The key roles at Ein Harod, as in other settlements of the kibbutz type, were secretary of the settlement in charge of all economic affairs and all affairs of major importance in the kibbutz (the so-called prime minister of the kibbutz); internal secretary in charge of social affairs; treasurer; purchasing officer; two work coordinators, who determined job allocation and the transfer of personnel from one branch to another; and two work organizers in charge of the daily work rosters of the settlement.

However, all routine matters and those related to policymaking were dealt with by the large secretariat, composed of the eight administrators and twenty additional members without executive functions. This body, most of whom had management experience, constituted the actual authority of the kibbutz. It settled all important problems, bringing to the general assembly only matters of basic principle and questions on which it could not agree. The large secretariat also convened the assembly and decided on its agenda.

In addition to these three bodies, there were also a number of committees active in various spheres of the settlement's social life. These committees made recommendations to the secretariat as to plans, projects, and arrangements affecting their area of competence. A number of them, though, enjoyed considerable freedom of action and independence in their sphere of responsibility.

The most important committees were the education committee, having two subsections, one to deal with infants and the other for school-age children, the cultural committee, and the health committee. Ein Harod did not introduce economic committees, and in this it differed from the practice adopted by other kibbutzim. The active secretariat acted as the supreme body in these affairs, and all policy decisions regarding the economic set-up and their implementation were concentrated in it. Such centralization did not exist in social affairs, and it is in them that the committees fulfilled many varied roles, although even here their scope was more limited than it usually is in most kibbutzim.

Election of members to various roles of primary or secondary importance was carried out through various arrangements. Theoretically, it was the general assembly that had supreme authority and legitimated élite roles and their power, but in practice, most officeholders were not directly elected by the assembly. Rather, the assembly appointed an electoral committee, which contacted various members and enquired as to their readiness to accept positions on various committees. The list of candidates was submitted for formal approval to the general assembly, which performed in most cases a ratifying function and not an electoral one. Members of the large secretariat, on the other hand, were elected by secret ballot by all members of the kibbutz.

A small group of members, holding the key positions and led by the settlement secretary, had consolidated their authority over all important areas of settlement life. Even matters not directly subject to the control of the active secretariat, and dealt with by various committees, were subject to its approval. This small group was not entirely independent, but was closely linked to the large secretariat, an effective instrument because of its limited size and the close contacts it maintained with key position holders. Thus, the large secretariat was in practice the supreme governmental instrument, but its basis of legitimacy lay with the general assembly. This assembly abandoned, indeed, all day-to-day administrative tasks when the processes of specialization and differentiation became more evident; however, it continued to perform important communicative functions in maintaining social integration. It further made information concerning events in the settlement available to rank and file, assuring the ordinary member's interest in affairs outside his own restricted personal field. The assembly also preserved vital contact between the élite and the rank and file.

Functions of élite control were also part of the assembly's role, since leadership was here subject to review and criticism on the part of the membership. It appears that this function became less effective, for differentiation and role specialization were such that ordinary members were less able to recognize and judge the problems arising out of the management of a large settlement. In spite of this process of centralization, however, political structure continued to preserve a considerable degree of democracy, in accordance with the settlement's basic principles and the need of integrating all members in the collective effort.

In addition to the assembly itself, several other mechanisms maintained the effectiveness of democracy and the efficiency of administration. The electoral practice was thus geared to assure the nomination of members who were qualified and who commanded general support. Of particular integrative importance here was the elimination of candidates who, while supported by the majority, were likely to engender hostility. Another principle was that of role-rotation: with the exception of special cases, élite role holders did not serve on administrative pools for more than one term of office (three years for the secretary). On the other hand, much of the potential loss to specialization was prevented by electing past functionaries to the large secretariat and so putting their experience at the disposal of the smaller active body. The structure of the large secretariat also allowed new people to be introduced into the supreme governing institution. Another channel which ensured maximum participation by kibbutz members, though in roles of lesser standing, was committee membership. Since a large number of committees operated for limited periods, almost all the members held one or another position every few years, both increasing integration and diffusing power.

A fundamental characteristic of the political system of the kibbutz was its high degree of intervention in most areas of community and personal life. This was in accordance with the basic collective and communal principles, according to which ideological commitment and social control alike extend to most aspects of life. However, while the authority of the community over the member was great, there was in practice little personal or group power. First, the holders of élite positions had no special material rewards and could not accumulate anything beyond the period of office. There was, of course, a possibility of creating a personal following; such following, however, was usually based on qualities that were considered desirable for the community. In any case, since the public did not tend to re-elect the same individuals to central positions, nor the incumbents often present themselves for re-election, there was no tendency for the formation of a formal political élite or for concentration of power in the hands of a small group.

By contrast to its scope of interaction, as a voluntary body, the political system of the kibbutz had no formal or legal sanctions. Unlike the moshav, where, as we shall see, such political vacuum created grave problems, here mechanisms of social control, functioning outside the political sphere, were highly successful. One of these certainly was social pressure, which could lead up to social ostracism. This can be a very powerful weapon in a society in which the individual is dependent to such a large extent on his fellow members; and although the large kibbutz was much less closely knit than the small kvutza, no member could survive in it when sent to Coventry. Most important, however, was the ideological commitment itself, consistently reinforced by powerful communicative and socializing mechanisms, calculated to indoctrinate and at the same time to create opportunities for individual fulfilment.

Recognition of the importance of education, and Ein Harod's belief that it could make a special contribution in this field, led the kibbutz to demand a separate school for itself. This meant a conflict with the movement, since the usual practice was then to establish a school in common with several other settlements. Also revolutionary was the demand to introduce twelve years of schooling,[31] when the accepted kibbutz opinion in the pre-State era was that extended formal studies were unnecessary for the development of personal social values; on the contrary, too lengthy a period might cause the youth to doubt the ideals of labor and equality and could produce deproletariatization and careerism and even lead to the leaving of the kibbutz. Ein Harod, while aware of the dangers, understood that the youth could not fulfil élite roles if they lagged behind in intellectual achievement and in the capacity for independent and scientific thought. We cannot, of course, enter here into a discussion of the problems of collective education in general and of the merits or drawbacks of Ein Harod's specific solution [32] (since adopted, by the way, in most kibbutzim). It is the innovation itself, however, which is significant.

Ein Harod also encouraged the members to lead a rich cultural life. As was seen above, the large kvutza strove to eliminate differences between intellectual and laborer, and to show that within the frameworks of its dynamic and varied society, conditions would arise for achieving an integrated and a whole personality. The personal qualities of those who founded Ein Harod were undoubtedly suitable for this ideal. In addition to the general emphasis on culture and spiritual values, there was a large number of members talented in literature and art, for whom creative expression was a personal necessity. These gifted members found Ein Harod conducive to creative activity; the community valued their work and was proud of their accomplishments, and facilities were made available for them. Ein Harod was the first, and for a considerable time the only, settlement that allowed creative artists time off each week from their regular work to concentrate on their art. The special conditions at Ein Harod and the presence of gifted members in turn produced a special atmosphere of cultural alertness in the settlement.[33] The most important enterprise in the artistic sphere was the establishment of the Artist's House. This served as a museum where the best productions of local and other kibbutzim artists were exhibited. The House converted Ein Harod into an artistic center for the whole region and for the kibbutz movement. It encouraged artistic production, in both painting and sculpture, and inculcated aesthetic values into the settlement children.

Another integrative and educational mechanism consisted of various seminars and long- and short-term courses, ranging from economics and professional matters to ideological and political subjects. These activities involved a considerable part of the settlement's members at one time or another and served also to strengthen the contacts between rank and file and the movement hierarchy.

Ein Harod also pioneered in reviving Jewish traditional elements. Originally, the attitude of the kibbutz to the traditional Jewish way of life, and especially to religious values and rituals, was highly negative, as was the orientation of all secular collective movements.[34] In time, however, Ein Harod realized that the task of creating a new national culture demanded a degree of continuity, and that this continuity could, in a modified form, enrich the kibbutz itself. Accordingly, traditional festivals were revived but given a new form and substance, emphasizing national and agricultural values, and minimizing or ignoring purely religious features.[35]

In the preceding pages we have analyzed some of the basic problems and processes of Ein Harod, mainly in terms of its ideological and institutional pattern. Because of insufficient data and limited space, we could not document many important events, and an unfortunate impression may have been formed of a smooth resolution, within this pattern, of most of the contradictions and conflicts mentioned. This, of course, is far from true, in several important ways.

Many of the institutional solutions reached were accompanied by serious, even agonizing ideological strains, disputes, and divisions, of which we specified only those that actually split the community. Moreover, they often were resolved by merely temporary solutions and were destined to erupt again. Some of the contradictions resisted resolution, embodying as they did conflicting principles. *It is precisely the genius of the kibbutz, however, that it can live and thrive with seemingly irreconcilable contradictions.* This genius manifested itself in Ein Harod in a number of ways. It developed, first, several informal mechanisms of secondary institutionalization, which minimized, explained away, or isolated the gap. Among these mechanisms mention can chiefly be made of two: rationalization, the representation of change in terms of the existing ideology; and the strengthening of ritual and symbols representing basic values, and their adjustment to novel forms of behavior, thus making it possible to develop a new framework in which the older values are symbolically maintained.

These mechanisms were developed here precisely because there was an intensive attachment to values, which compelled a solution; and this was made possible by the fact that in the kibbutz the extent of initial institutionalization and formalization of norms was relatively small, having developed, much more than in the moshav, in a trial-and-error way. As a result, Ein Harod could maintain the original ideology side by side with new behavior patterns and criteria of day-to-day evaluation. In other words, the gap between ideology and reality was often itself *institutionalized rather than bridged*, especially with the help of mechanisms that tended to blur symbolically the specialization and the differentiation that developed in reality. To give an example, the role-differentiation among specialists, between specialists and services men, and between élite members and the rank and file is to some extent "adjusted" to the original principle of equality by means of symbolic participation in lowly roles (such as a one-week tour of kitchen duty once a year, performed alike by the cabinet minister who is a member, by the secretary, and by the branch manager).

However, the clearest asset of Ein Harod, in terms of sustained change and development, is its entrepreneurial and innovating drive, which absorbs and initiates change. The greatest scope for initiative was in the economic sphere, and especially that of production, where with the exception of hired labor, there were few ideological restrictions on "rational" behavior. Social entrepreneurship was more delimited—but here too the range of invention, experimentation, and flexibility was considerable. As we saw, this phenomenon was motivated by the basic ideology of the kibbutz and supported structurally by the constant recruitment of new social forces and the circulation of élites. How successful it really was, we shall only be able to see fully when we discuss Ein Harod's child—the Hakibbutz Hameuhad movement. Already, though, the scope and success of its internal and external activities, and of its flexibility in the face of crisis and disbelief, makes an impressive record.

The history of Ein Harod, as told here, ends abruptly in 1948, but its story, and the story of the kibbutz in general, goes on. The passage of time has brought many changes to the collective form as a whole, under the influence of both internal and external factors.[36] Internally, there was numerical growth, in the course of which the initial nuclei often grew from a few dozen members to several hundred. This growth was closely related to economic development. Secondly, there was the process of maturation of individual members, associated with ageing and with the increased salience and problems of the family and of child-rearing.

Beyond the general demographic changes, the continuous existence of the collectives as communities spanning into new generations has entailed the routinization of the ideological drive, the formalization and differentiation of roles, and the development of various social subsystems. The original homogeneity was weakened and altered by the growing division of labor, by the articulation of the authority structure, and by the crystallization of various solidary groups. Concomitantly, distinct élites emerged within the originally classless setting, such as production managers, technical experts, and professional political leaders.

Externally, there was a clear change in the relative standing of the kibbutz in the broader society after the establishment of the State of Israel. During the Yishuv period, the kibbutzim had the criteria of high status: public activity, length of residence in the country, and rural as well as national pioneering; as a result, the member of the collective had enjoyed an élite position. With the establishment of the State, new values and élites have emerged, embodied chiefly in the professional, governmental, and entrepreneurial spheres. This has influenced the standing of the kibbutz and the kibbutz-member in a number of ways. On the one hand, the values symbolized by the collective were affected: some of them, such as equality and simplicity, diminished in general importance and salience; in others, notably pioneering and national service, the collectives lost their previous monopoly to formal national organizations, such as military élite corps. The concrete position of the kibbutz in society decreased: it was now becoming dependent upon the State for the provision of many resources and lost much of its bargaining power vis-à-vis other organizations.

This loss in functions, status, and power weakened the original basis of identification with the kibbutz and caused, in some cases, a search for avenues of mobility. It often put the ideologically more devoted members on the defensive and reinforced the economic trends toward novel spheres of enterprise and activity, emphasizing education, specialization, and diversification and creating new élite positions.

As we saw, however, in many of these developments Ein Harod had actually anticipated trends articulated later and had shown, albeit under different conditions and stimuli, the capacity and mechanisms for meeting and leading the challenge of social change. Thus, while historically we stop in midstream—as no good story should—justice is essentially done to the basic characteristics and problems.

4

A Moshav: Nahalal

The following case study deals with the last of the major settlement types established by the Jewish community in Palestine. The fact that the moshav was a relative latecomer upon the colonizing scene, emerging into a somewhat crystallized institutional setting and capable of drawing upon previous experience and of selecting from existing patterns, was certainly of great significance for its development and is the point of departure in our story. However, it also means that the period from establishment to 1948 covers essentially one generation only and primarily represents the initial phase of creation and institutionalization. In this phase, many of the problems and strains that are built into the pattern and that emerge forcibly when the moshav comes of age are but latent or embryonic. It will be recalled that this was to some extent the case with the collectives, where the post-independence era brought out and intensified a host of challenges. There is, however, an important distinction between the two: the earlier history of Ein Harod actually anticipated, as we saw, most of these challenges, especially such internal processes as maturation, demographic growth, and social and economic differentiation; and the analysis of their impact, and of responses made to it, during the pre-independence period gives us an insight into problems and adjustments that were to become common later. Not so as regards Nahalal and the moshav in general; cutting off the story in 1948 would mean a one-sided and perhaps even a misleading picture. It is for this reason that here the presentation is, in certain aspects, carried beyond the chronological limit set.

The data are organized around two themes related to institutional characteristics: the process of genesis and crystallization, and the pains of change. In other words, it is not stages of economic or community development that mainly serve as the basis of the division, but rather distinctions in the nature and the problems of the social institutions of the moshav. The cutting point selected here was the middle thirties, which occasioned, as will be seen, the first major attack on a basic institutional principle of the smallholder's cooperative—independence from hired labor. As elsewhere, institutional change in Nahalal was gradual and unevenly spread over various spheres, rather than sudden and comprehensive; the dividing line chosen is thus necessarily arbitrary and artificial. Institutionalization is a continuous process, and we try to show the internal seeds of this continuity in our analysis. At the same time, the nature of this process did change significantly, and the division is thus useful in pinpointing the new problems that then emerged.

The First Phase—Birth and Crystallization

Theodor Herzl, the father of political Zionism, envisioned the colonization of Palestine in terms of "cooperative communities, productive cooperatives that include not only plantations, but are built upon mixed farming. They produce primarily for their own needs and market only the surplus, through a joint cooperative organization." [1] The idea of a smallholders' cooperative village was thus present in the early guiding image of Jewish settlement; its realization in practice, however, came much later and only after the colony and the collective had been established. Indeed, the first actual proponents of the moshav did not assume the mantle of heirs to the original vision; rather, they acted in response to their pressing personal problems, and in reference to the colonizing experience by then accumulated in the country. Initiative for the innovation lay, in fact, with a group of Second

Aliya immigrants who had been eking out a precarious existence as hired agricultural laborers in the colonies and had been weighing the best way of becoming farmers and settlers in their own right. They decided that both the private and the collective patterns, while embodying much that was valuable and successfully tested, failed as wholes to answer certain basic social and economic needs, individual as well as national.

The main agrarian drawback of the moshavot was thus seen in the susceptibility of monoculture to crises, the plantation colonies producing for unstable foreign markets, and the field-crop colonies suffering from the frequent droughts common to the area. As was seen in the case study of Petach Tikva, the war years sharply aggravated this factor, and its impact was felt even more since dependence upon external conditions clashed with the professed ideal of the self-reliant farmer, master of his fate. The war also underlined the importance of strong community organization and cooperation in times of stress. While it was noted that places like Petach Tikva did succeed in rising to the occasion, this was considered a unique effort, beyond the fundamentally individualistic institutional terms of the colony.

The colony, incorporating as it did hired labor and free sale and accumulation of land, was condemned as leading inevitably to the *kulak* system, to speculation in and alienation of agricultural units, and to the general reversal of the process of social productivization and rebirth. In this respect, the fathers of the moshav spoke with the voice characteristic of the pioneering and socialist immigration as a whole, identifying themselves with the founders of the kvutza. Not so, however, in other areas. Disagreement was voiced, first of all, regarding farm structure, since the early kvutzot, like the colonies, were based on monoculture and thus exposed to similar criticisms. Indeed, their extensive field crops required an inordinate amount of seasonal work, which the critics believed to be beyond their internal manpower capacities, and, they maintained, would ultimately lead to the employment of hired workers. Pioneering and productive ideology notwithstanding, such collectives, it was thought, were bound to lose their *Gemeinschaft* atmosphere and egalitarian character, and to turn into large business enterprises with employers and employees.

In any case, the kvutza pattern was considered incapable of creating strong bonds between the individual and the soil and of giving adequate scope to his initiative and energy. As was seen above, the idea of the kvutza was during that period generally fighting for recognition; the moshav founders thus reflected the widely held belief that it would not stand the test of time. A special and a more personal element in this negative evaluation was the care and education of children. The prospective settlers described were mostly people with established families; they were suspicious of communal encroachment upon the domestic domain and unwilling to risk the sacrifice of family ties for an experiment whose success was by no means assured.

Clearly, then, the ideal solution seemed to lie somewhere between the two extremes previously attempted, and in an effort to incorporate the family farm of the colony with the pioneering values and strong community solidarity and cooperation of the collective. However, the emergent pattern was by no means an artificial creation, a compromise based on selection and modification from existing features; a completely new form was built, with a life of its own, commanding devotion and enthusiasm. At the same time, the fact that the moshav had a long gestation and was evolved in a process of careful deliberation rather than in a flash of vision had a far-reaching significance for its subsequent institutional development. Before taking up this point more fully, though, the pattern itself requires a careful explanation.

Three major strands seem to have been woven into the basic image of the moshav:

1. *The common goal of all the pioneering groups: the threefold ideal of productivization, of nonexploitation of others, and of national service.* Concretely, this ideal was embodied in the following: national ownership and nonalienation of land, which was considered, it will be recalled, as a value in itself and the

basis of a new way of life rather than as a marketable commodity. "By founding moshavim," said one of the economic and social theorists of the movement, "we enable people to live on the land, and at the same time we put the soil—which is the origin of life—in secure hands. As the land will not become the individuals' property, they will not trade with it"; [2] performance of national tasks; public financing, planning, and guidance of the settlement, which were considered a condition for accelerated development and growth, but which primarily symbolized and safeguarded the mutual bond and responsibility between farmer and society; emphasis on own and on manual labor; and simplicity of life.

2. *The creation of a family-farm culture*, in the sense of the rural family life being truly dedicated to and organized around the farm enterprise. "Nahalal is naturally built on the basis of the family," says S. Dayan, one of the founders, in his reflections. "Its roots are the lives of its families." [3]

For generations, Jews have been removed from nature and agriculture and the hardships of farming; they tend to avoid it. Our difficult task is to attach every individual to agriculture by giving him a piece of land and telling him: Struggle with this piece of land, invest all your physical and mental powers in it. Think over each step you take and earn your daily bread. Do not hide behind the public's back.[4]

Again, A. D. Gordon, the philosopher of the Jewish productivization movement:

If you want a Kvutza or any other form of settlement, do not consider as home a place in which you are kept squeezed like herrings, thinking that you manage wonderfully—people are not herrings and they cannot be kept in a barrel. There is dynamism among persons and life and a whole world.[5]

In this context, the general pioneering ideal of self-labor, mentioned above, is colored with the additional tinge of reinforcing family tenure, management, and cultivation:

Jewish agriculture in this country, with its low crops and high expenses, can be carried on if suitable farming methods are applied and if the farmer does not strive for those nice things of life that certain classes in our nation desire. The farm that should be aimed for is one which excludes all hired labor and is managed by its owner and for him: self-supply is its backbone in most of the country's regions, and self labor is its unbreakable law.[6]

The viability of the family farm is also the avowed meaning of the values guiding the character of the projected land system: the emphasis on the smallholding size of the unit, manageable in terms of the household; the emphasis on the farm as the main (if not the only) source of employment, but also the main source of income; the value of doing one's best by the land and of achieving the utmost with the farm's means of production; and the stress on mixed farming and production for own consumption, expressed not only in terms of economic security and of an even production curve, but as a symbol of autarchy and independence, and of the all-round farmer bound to his farm by broad and diffuse ties. E. Yaffe, the most prominent founding father and the author of the moshav program, says:

The mixed farm is almost independent of external markets and thus it spares the farmer many problems: (a) Everyone who depends on the market becomes perforce a merchant whose mind and heart tremble like the prices in the market. He is constantly preoccupied with the markets and cannot strike roots into the soil; (b) The basic mixed farm furnishes the farmer with all his needs; it captures his heart and binds him to his farm with unbreakable ties; (c) The farm that is dependent on the market is not stable. Its farmer is constantly threatened by losses, his nerve weakens; (d) The basic farm produces food over and beyond the market-dependent farms in forms that a family needs; (e) By producing for the market the soil becomes abused and finally impoverished; (f) We need the basic farm which provides the farmer with a livelihood out of his own work, and also in order to achieve a national and social goal. It is the Jewish farmer's task to free the agriculturist from his economic subserviency and lift him from his cultural mediocrity.[7]

3. *The establishment of a* Gemeinschaft. The image of the community as a family of families embraced social, stratificational, and organizational aims:

(a) a high degree of social interaction, solidarity, and mutual aid;

(b) a basic homogeneity and lack of differentiation, embodying occupational and farm structure, public responsibility and functions, and income and style of life. However, in contrast to the collective, there was to be here no mechanical equality, but rather the equality of life chances (safeguarded by the equitable distribution of the means of production). Thus, the community would recognize the value of individual drive and skill, representing a national as well as a family-farm asset, and it would only oppose a clear polarization on the one hand, and conspicuous consumption on the other;

(c) a strong organizational and cooperative structure, binding the community and providing economic organs, services, and a framework for corporate decision-making.

Like most pioneers of the Second and the Third Aliya, the settlers of Nahalal were essentially secular in their aims and outlook, skeptical of religious truth, and opposed to religious education and way of life. True, they were in this, as in many other spheres, less extreme and militantly active; they lacked the missionary secularism present in the Labor Battalion (and many collective nuclei), while the very salience for them of family and household made for some maintenance, if not for actual ceremonial observance, of fêtes and festivals significant for the home; indeed, Nahalal, unlike Ein Harod, built its own synagogue. At the same time, the founding ideology of Nahalal—similar to that of Ein Harod and totally unlike that of Petach Tikva—represented national revival and social transformation and not any traditional regeneration.

So much, then, for the bare ideological facts. However, even a cursory analysis of the value system described shows that its elements, while reinforcing each other, also formed patterns of potential inconsistencies. Such inconsistencies, competition, or strains were thus implicit in the three distinct strands mentioned —the pioneering, the familial, and the communal. To some extent, the situation was similar to that described in Ein Harod, whose early history saw the articulation of the built-in conflict between national and settlement goals.

However, in the patterns of both the collective and the colony, a clear decision is made as to the relative dominance of the private and the community spheres. Not so in the moshav, where the normative scheme provides for mutual limitations and support, but does not determine which sphere is to be supreme. A much more basic strain, though, seems to lie in the fact that the image of the moshav member represented two fundamentally contradictory characteristics and requirements, that of being a *peasant*, and that of being *modern*. For surely these are the essential dimensions of a set of expectations in which, on the one hand, there are noneconomic limitations on the use and the cultivation of land, on the size and structure of the enterprise, on production for the market, on the free use of labor, and on the spending of income; and on the other, premium is put on enterprise, use of advanced techniques and tools of cultivation, and on modern cooperation and government. The noneconomic farmer is, of course, also the creature of the collective. The special limitations of size, specialization, and differentiation seem, however, to be particularly characteristic of what we called the culture of the family farm, or more properly, the culture of the small holding.

Of course, the contradictory image of the "modern peasant" was initially not only an asset in the existing situation, but a necessity in the context of the social expectations and characteristics of the settlers to come. It was, in fact—to paraphrase the American dictum—an ideology of the people, by the people, and for the people.

Be that as it may, however, the strains discussed are built into the moshav pattern, and most of the processes and problems analyzed later on can be traced and understood in terms of these basic dimensions of the original image.

Before proceeding now with the story of the actual articulation of the moshav image in the village of Nahalal, one further remark must be made on the significance of this background for future change. As was mentioned earlier, the creation of the moshav pattern was a relatively long and leisurely process, conditions of war preventing an earlier implementation, in spite of the mounting social and economic pressure upon the agricultural workers in the colonies. This enabled the founding fathers of the movement to forge themselves into a group. No less important, it gave them time and opportunity to translate the basic guiding image into a systematically thought out and elaborated set of concrete rules and decisions.

This chance was further enhanced by the existence, for several years, of a type of "pre-moshav" in the form of small settlements established near the large colonies in order to give the hired agricultural laborer an auxiliary farm attached to his house, and so to increase his income and independence. These pre-moshavs were considered partial and temporary solutions and of necessity incorporated only some of the institutional arrangements envisaged for the moshav proper. At the same time, they provided a valuable testing ground and a running-in period. Nahalal could thus become, from its beginnings, a smoothly running concern, with provisions and solutions (if not resources) actually anticipating problems. However, the detailed formalization of the institutional structure (soon to be documented) potentially limited the flexibility of the system later on. In this respect, the lesser initial crystallization of both Petach Tikva and Ein Harod, and their necessarily greater reliance upon trial and error, gave them, perhaps, a future advantage in structural innovation and change.

As has been seen in the case study of Petach Tikva, the First World War made the situation of the agricultural laborers in the colonies desperate. While the war suspended actual preparations for settlement, they were vigorously renewed immediately after the cessation of hostilities. Since some members of the original nucleus had died or dispersed, reinforcements were recruited among like-minded people from the newly arrived Third Aliyah. On the morning of September 11, 1921, the first twenty families of Nahalal pitched their tents on a hill in the Jezreel Valley. Soon after, additional families joined the advance party, bringing the total to seventy-five households in all, mostly members of Hapoel Hatzair party.

In certain important aspects the settlement of Nahalal was carried out in more favorable conditions than that of Petach Tikva or Ein Harod. Chief among these were the generous tract of land made available by the Jewish National Fund, and the presence upon the scene of the newly established colonizing instrument, the Foundation Fund. Thus, while the necessary capital was provided piecemeal, there was the formal commitment of organized public backing. On the other hand, the Jezreel Valley, though potentially fertile, was then a swampland swarming with malaria-carrying mosquitoes. There were no proper roads, no means of transportation, and no centers of population and services nearby. The settlers, moreover, were no longer very young, with more than half of the men over thirty and with the burden and the responsibility of families. In this respect, therefore, the initial years were here, as in the other villages, a backbreaking and a heartbreaking period of unremitting struggle, of pioneering in the true sense of the word. At the same time, the moshav was spared the crisis of continuity that Petach Tikva suffered, as well as the social cleavage between the main social groups in Ein Harod. As soon as the physical obstacles were overcome, farm and community building could begin.

As was mentioned above, the most striking characteristic of this process was the early institutionalization of the basic guiding image into a detailed agrarian, social, and organizational blueprint that, with some exceptions, sprang as it were fully armed, like Athena from Zeus's head.

The first focus of thought and internal "legislation" in Nahalal was the translation of the ideal of the family-farm culture into concrete and viable arrangements. In this respect, the period of ground-breaking was inevitably a moratorium, since the drainage of the swamps not only required the settlers to work in a communal group, but also necessitated the employment of hired labor. By 1925, however, the preparatory work was largely completed; the drastic reduction in the number of the temporary laborers (from 181 in 1922 to 68 in 1925)[8] signified the institutional birth of the moshav.

The basic unit of production adopted and given equitably to all founding families was an area of twenty-five acres. In the beginning, the household maintained itself by outside employment. This, though, was to be a temporary measure, as brief as possible, and was to disappear completely with agricultural intensification, and not to constitute a competing or an alternative activity. The plan envisaged progressive development, with scope and intensification of production increasing in proportion to the size, the requirements, and the resources of the household. Already, in 1919, the preparatory committee had formalized this principle for future use:

> The plan is designed for settlement on naturally fertile soil or soil that has been improved by manure. The farm will not be intensive at the beginning: it will gradually become so as the settler's family and thus the available manpower grows. This plan presupposes that the settler will manage the farm himself, and—aided, if necessary, by members of his family—will carry out the whole year's labor.[9]

It was nevertheless recognized that the area provided, when intensified, might exceed the manpower potential of even two generations. Indeed, the twenty-five acres were allocated because a large parcel of land had just been bought in the Jezreel Valley.[10] There was, consequently, the basic anticipation that all land found to be beyond the capacity of the farm-family manpower, and thus requiring hired labor for cultivation, would be returned and put at the disposal of new settlers. To maintain the integrity of the self-contained family farm, there was an absolute ban on individual transfers of land, as well as a clear provision for the indivisibility of the farm inheritance, thus preventing fragmentation.

The tenure agreement between the National Fund and each farmer formalized these two basic principles. Otherwise, the tenure was completely secure and singularly free of conditions. Its term was forty-nine years, with largely token payment, and was almost automatically renewable. There were, in fact, almost no legal teeth in it, and in particular no provisions for terminating and controlling unproductive, mismanaged, or abused units or of preventing subleases. Most of the controls for the various limitations and provisions of the blueprint actually lay in voluntary ideological commitment, in informal social pressure, and in the dependence upon village institutions. To this crucial circumstance we shall have occasion to return again and again.

It will be recalled that the image of moshav society was essentially one of a community of families, bound together by high solidarity and by an intensive network of interaction. A basic lack of differentiation—occupational, social, and economic—was considered an essential aspect of such community integration. As will be clear from the preceding discussion, this function was institutionally safeguarded by the agrarian arrangements, which defined and protected family farming and meant that, with the exception of a few nonagricultural families necessary for some specialized services, all the households would have an essentially similar farm character and basis.

It was, of course, realized that the very same farm system, with its emphasis on progressive intensification according to differential needs and resources, was bound to create, in the course of time, considerable individual variations; but while the image of the good society did not imply any mechanical equality or uniformity, it nevertheless precluded extreme polarization. Two institutional mechanisms were calculated to control such a crystallization: on the one hand, provisions were made to assure a minimal standard of living, either by mutual help and corporate incentives to farms that ran into trouble, or by the allocation of public work and financial help to hardship cases; and on the other, constraints were enacted on accumulation of property and on outside investment. Again, the main safeguard was informal social control and ideological identification.

Since corporate action and responsibility was a fundamental principle of the moshav, careful thought was given to the articulation of the organizational framework of the village, which would combine direct democratic government with efficiency in the provision of modern services. The supreme authority over all the public aspects of the community was vested in the general assembly of adult members, which functioned as a legislative body. It met once a year, deliberated and decided matters of general policy, and elected, by universal and secret ballot, the executive officers of the moshav—taking care that no member was chosen to more than one office and that the principles of rotation and of maximum public participation were maintained. However, while the supreme body was of necessity a single one, the administrative offices themselves were divided into two spheres of activity, organizationally and substantively distinct: the social-municipal and the economic-cooperative.

The village council (a little later with a selected executive committee) was the framework charged with the first of the functions mentioned, that is with social and cultural life, with the provision of basic services, and with social control. Organized and active from the very beginning, it was helped in its work by a secretariat and, as the need arose, by special elected committees in such areas as education, health, security, and recreation and culture. This combination of a basic blueprint with provisions of flexibility is exemplified in the area of mutual aid. It had been originally thought that such aid, to farms and households alike, would be entrusted to voluntary action by friends and neighbors, both reflecting and strengthening the high community solidarity. It was soon found, though, that spontaneous aid was neither equitable nor efficient, and it was formally handed over to the community, through a newly created committee on individual and welfare problems.

The economic-cooperative structure, by contrast, was a relative latecomer, for the obvious reason that in the embryonic farm and marketing situation its actual functions were limited. The institutional blueprint of the cooperative organization, however, was spelled out clearly and unequivocally from the beginning. The basic principle of this outline was that cooperation was an inseparable part of the community fabric: not just a means towards achieving economic ends, but rather a facet in the solidarity and the integration of the village. For this reason, a centralized and comprehensive cooperative structure was envisaged: it would be directed by a joint board or committee and assisted by specialized branch committees. Within one multipurpose framework, it would administer supplies, marketing, credit, and cultivation of any cooperative branches. Thus, while the actual implementation came piecemeal in response to concrete needs, and while specific organs and details would be changed and modified for purposes of efficiency, the essential pattern was considered a fundamental arrangement.

The actual impetus to cooperative organization was given by the growth of the dairy and the poultry branches, when the members of the moshav had to solve the problems connected with storage, with the marketing of excess produce, and with credit facilities to bridge gaps between deliveries and payment. In 1926 marketing was entrusted to the communal storehouse, which earlier had only the function of centrally buying farm and household supplies. At the same time a special credit institution, the Farmer's Credit, was established. Still later, with the further development of the poultry and dairy branches, the expansion of the drained area, and the rise in turnover capital due to the increase of marketed produce,—it became necessary to separate marketing and the supply of products for the farm from the communal storehouse. In April, 1927, a special cooperative branch, Anvah-Nahalal, was founded for this purpose, and the function of the communal storehouse was again limited to the supply of household goods and some items of farm production.

Again, some years later, it became obvious that there was no social or economic point in maintaining a separate credit institution, since there was a direct connection between the economic situation of each member and the credit granted him on the basis of mortgaging his products. So in 1936 the Farmer's Credit and Anvah amalgamated and were henceforth called Anvah—a cooperative agricultural society for marketing, supply, and credit in Nahalal.

In the sphere of production, the intensification of land cultivation and the increased use of water brought about an increase in the use of agricultural machines. In 1935 the agricultural machines branch was established, and a tractor, combine, and other tools were purchased. When the amount of marketed product and of supplied goods grew, the need for transportation arose, and in 1935 trucks were purchased. In time, these and other activities were concentrated in Anvah, which became the sole cooperative organ of the community responsible for marketing (including field crops); purchase of supplies, seeds, and manure; the water economy; breeding; milling and grinding; transportation; a cooling house and packing house; instruction in the various farm branches such as plantations, poultry, etc; checking amount of milk for dairy reports; giving credit to members and receiving deposits; and agricultural machines.

Most of the specific institutional arrangements discussed are reflected and focused in the ecological layout of the village. Village ecology is planned and evaluated chiefly according to seven basic agricultural and social criteria: security; work distances; social activity distances; efficiency for administration and services; farmyard efficiency; scope for development, mechanization, and rationalization; and development and maintenance costs.

These aims cannot be fully accommodated in any one pattern, and it is usually a compromise between agrarian efficiency and community considerations. Thus, the traditional gathered village seems to be advantageous for an efficient defense system, for services and infrastructure, and for social interaction and solidarity; it leads, however, to an excessive fragmentation of land and creates great distances between farmyard and outlying land, impeding rational cultivation. By contrast, the scattered village has the obvious advantage of the settler's house being set within his fields for more efficient farming, at the price of poor services and a less intensive community life. The striking feature of the ecology of Nahalal is not only its careful attempt to create the

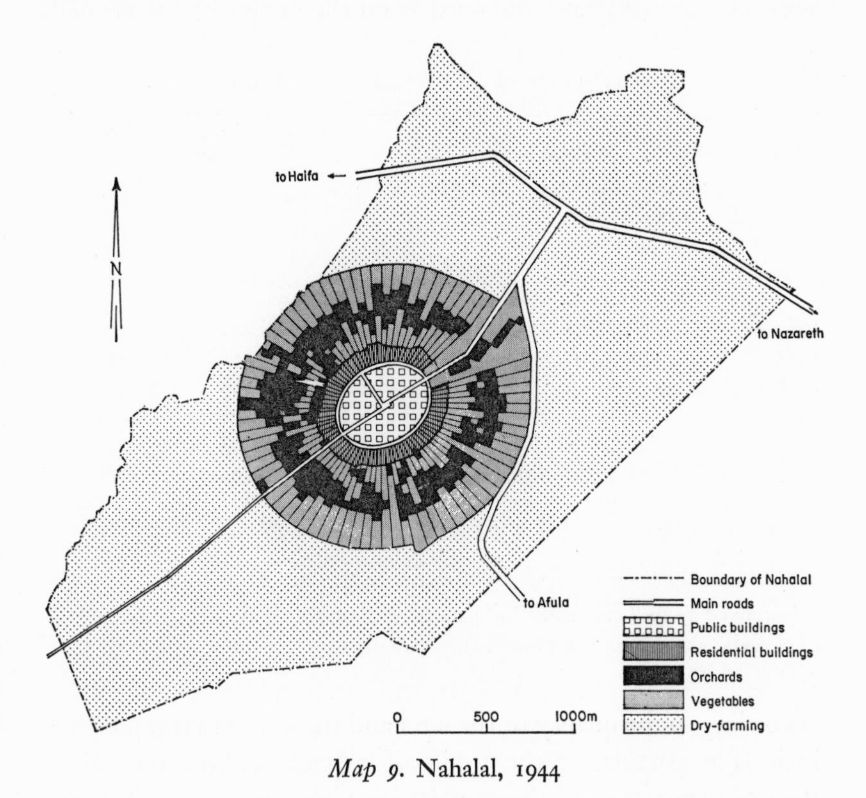

Map 9. Nahalal, 1944

"half-scattered" village, to combine the best features of both types; of importance also is the insistence upon maintaining a fair and equal land distribution.

As can be seen from the frontispiece and Map 9, of 1946 and 1944 respectively (which illuminate the essential character of the smallholder's cooperative perhaps better than any sociological analysis), the special shape of the "half-scattered" village in Nahalal is due to a concentric planning. In the heart of the circle, the various public buildings are placed. The area that encircles this core is divided into wedge-shaped plots of 2.5 acres each, and each settler's plot is attached to his house. The professionals and artisans among the settlers receive a plot of 1.25 acres each attached to their houses in the heart of the circle, but no other land. The bulk of the cultivated area, excepting the farmyard branches, radiates outward from the center and is divided

Table 17. Development of dairy and poultry branches, 1934–1939 (up to World War II)

	Average per farm			Total quantity	
Year	Return from milk, eggs, and poultry in £P	Eggs in units of 1,000	Milk in 1,000 litres	Eggs in units of 1,000	Milk in 1,000 litres
1934	66	4.5	5.2	341	392
1935	78	4.3	7.1	326	532
1936	115	6.9	10.2	499	735
1937	100	6.7	10.2	473	721
1938	99	5.8	10.2	408	717
1939	102	10.3	11.5	700	780
Average of the period	93	6.4	9.1	461	646

Compiled from J. Shapira, *Nahalal: Its Beginnings, Ways, and Achievements* (Tel Aviv: Am Oved and Tnuat Hamoshavim, 1947), p. 320.

according to the quality of the land and the kind of crop possible in it. The parcels are thus units of efficient cultivation rather than of formal tenure; they can support any given crop on large fields and facilitate the use of cooperative mechanization or of partial cooperative activities (in instruments, tools, etc.). Within this basic arrangement, however, (with the actual division modified according to requirements) care is taken to ensure that each settler has a similar number of plots, representing not only the same quality but, as much as possible, also the same formation and scatter.

So much for the agrarian and social arrangements institution-

Table 18. Returns from marketing of production according to branches and in relation to general income, 1934–1939

Branches	1934		1935		1936		1937		1938		1939	
	£P	%	£P	%	£P	%	£P	%	£P	%	£P	%
Field crops	2,213	20.0	2,065	16.4	1,806	12.0	1,120	8.5	824	6.8	585	4.2
Milk	3,817	34.5	4,920	39.2	6,363	42.2	6,258	47.4	4,964	41.2	5,540	39.6
Marketing cows and calves	2,104	19.0	2,074	16.5	1,684	11.2	1,162	8.8	2,028	16.8	2,150	15.4
Eggs	375	3.4	653	5.2	983	6.5	857	6.5	912	7.6	1,489	10.6
Poultry	598	5.4	237	1.9	239	1.5	170	1.3	526	4.4	427	3.0
Vegetables	369	3.3	369	2.9	973	6.5	556	4.2	1,003	8.3	1,738	12.4
Plantations	845	7.6	701	5.6	1,164	7.7	1,149	8.7	1,130	0.4	2,072	14.8
Miscellaneous	754	6.8	1,543	12.3	1,870	12.4	1,933	14.6	657	5.5	–	–
Total income of all farms	11,075	100.0	12,562	100.0	15,082	100.0	13,205	100.0	12,044	100.0	14,001	100.0
Average income of farm unit	152		172		206		196		165		193	

Compiled from J. Shapira, *op. cit.*, pp. 320–321.

alized during the period of consolidation. We must now examine, if only briefly, the actual processes of growth and development which took place within this framework.

Economically, the phase under consideration can be subdivided into two distinct periods. The first period (1921–1926) was one of financial consolidation; most of the basic investment schedule was completed, and the farms entered a cycle of regular production. The second period (1926–1936) was one of

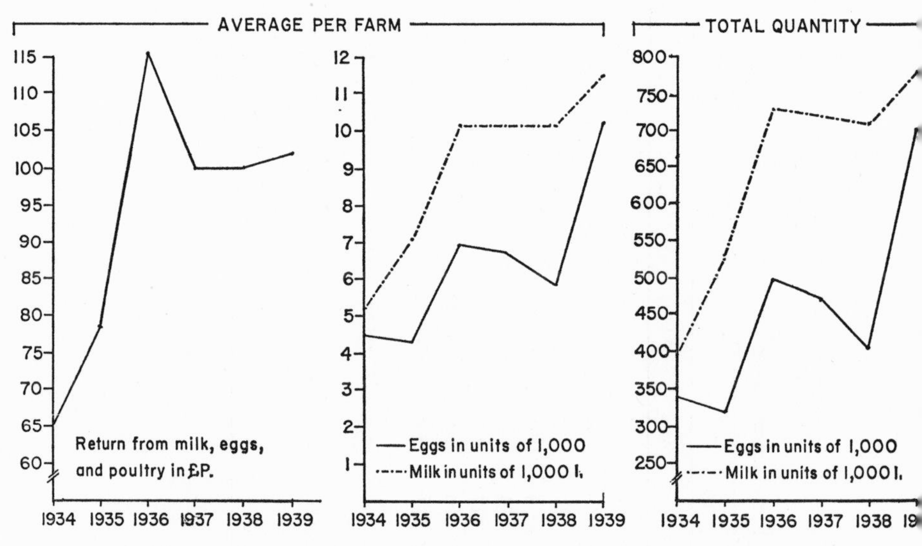

Chart 2. Development of dairy and poultry branches, 1934–1939

sustained development and prosperity. It seems that this process was characterized by a regular over-all increase in the produce and the income from individual farms and from branches, as well as by a balance between the crops making up the mixed economy. Tables 17 and 18 (and Charts 2 and 3) bring out this fact in respect to the years 1934–1939 (no dependable data being available for the earlier years).

A somewhat different picture is obtained in respect to demographic growth, in which there is a basic spurt and then stabilization. During the first five years almost the entire future complement is reached, the first two years representing the settle-

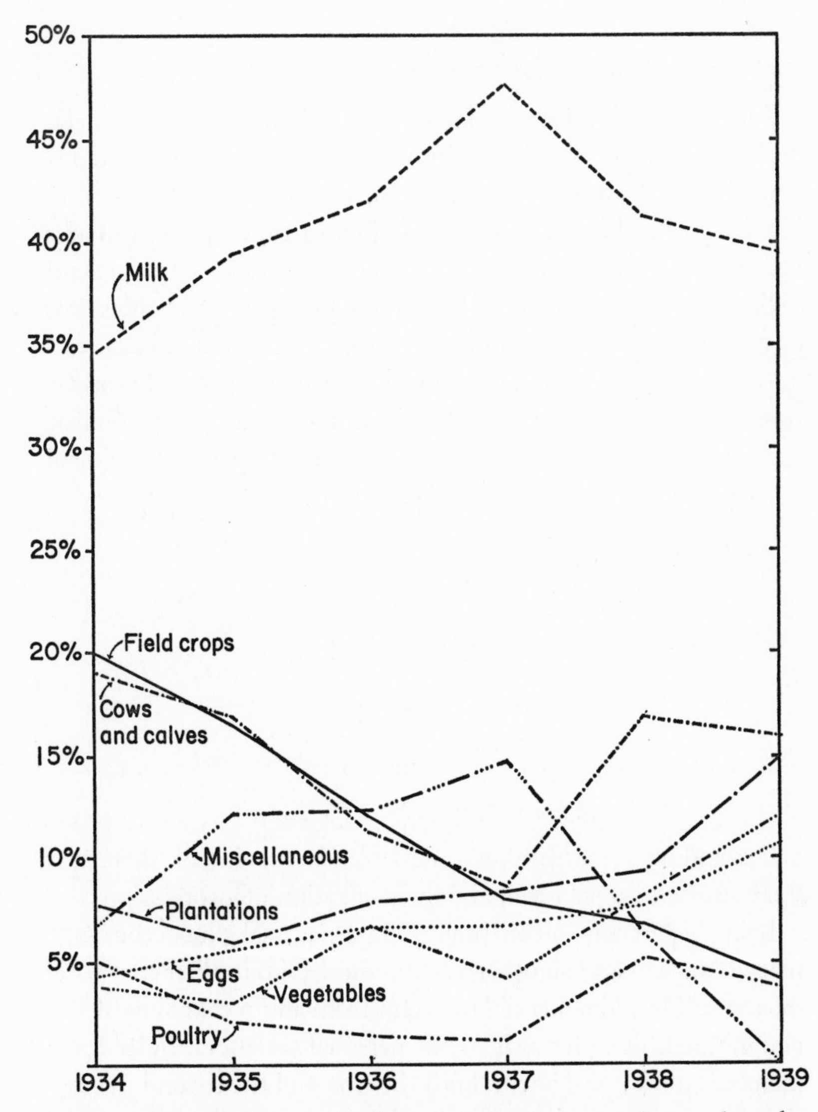

Chart 3. Returns from marketing of production according to branches and in relation to general income during the years 1934–1939 (in per cent)

ment of the population, and the following three the establishment of new families and the birth of children.[11] In comparison with the moshava and the kibbutz, the moshav is a relatively self-contained and closed social system. (See Table 19.)

The Second Phase—The Pains of Change

The factors of change, which began to impinge upon Nahalal in the thirties, but which were more fully articulated and intensified after 1948, were essentially those that we have described in relation to Ein Harod. These were, first of all, conflicting duties placed upon the village by national obligations and, later, general social and ideological shifts that affected the place and the status of rural settlement and pioneering on the one hand, and of voluntary organizations performing these functions on the other. Internally, there was the process of maturation and, in particular, the growth of a new generation less committed to the original

Table 19. Growth of population in Nahalal, 1921–1936

Year	No. of persons
1921 (advance party mainly)	30
1922 (main body)	365
1925	629
1931	674
1936	821

Compiled from J. Shapira, *op. cit.*, p. 220.

values and more susceptible to new trends. Inevitably, there was the increased scope, level, and intensification of production.

Being a partner—albeit junior—in a national pioneering élite placed Nahalal and all other veteran moshavim in their first basic dilemma. This dilemma did not stem from any weakening of this commitment in so far as it meant personal sacrifices. Indeed, the members of the settlement, both the first and the second generation, were among the earliest volunteers for all security and military duties during the Arab disturbances, in the Second World War, and in the Israeli Army in the War of Independence and subsequently. They also took upon themselves the burden of establishing, guiding, and providing resident instructor teams for the new immigrant villages, created in hundreds following the establishment of the State. The problem was that providing training and employment for refugees beginning to come from Germany in the thirties required the introduction of hired labor. The acceptance of this challenge, so directly at odds with the basic image of the family farmer, required an agonizing decision, reflected in long and sharp deliberations, especially as it was feared that hired labor would displace self-labor completely. The task of immigrant absorption finally won—on the clear proviso that it was a temporary emergency measure—and for three years Nahalal received hundreds of such workers.

This precedent pointed the way in other emergencies; recourse to hired labor was continued when insufficient family manpower was left on many farms because of military and other public service of Nahalal men during the forties. In this way a wedge of ten years' usage—though by no means of recognition—was driven. This fact was of great significance in the context of other processes favorable to hired labor, which now began to be felt: the growing scope of farming and of agricultural prosperity, particularly in response to the war; the population increase due to immigration; and the cutting off of Arab farm products following independence. Later on, the reinforcing effect of transition to specialized farming, with its uneven work curve, also began to exercise its pressure. In consequence, what had earlier been a duty and a necessity now became a privilege, an advantage to be eagerly grasped, especially by the more economic-minded second generation, now a significant segment of the population, and among the farmowners in particular.

The village bulletin, an accurate barometer of the atmosphere in Nahalal, deals with this question exhaustively and presents the different points of view. Little by little, it seems, the original reason for flexibility—absorption of immigration—was forgotten, and a relaxation of the principle was advocated for different reasons, especially economic ones. In 1956, for instance, at the general assembly, one of the members said:

> Self labor is indeed one of the basic principles of the moshav, but present circumstances differ from those that obtained when the moshav was established and the principles laid down. The fact is that

farm expenditure and local and general taxes are now so high that a sufficient living cannot be made from agriculture unless considerably intensified. In order to do this and increase the income, hired laborers must be employed, the members of the family alone being unable to carry out all the jobs. Hired labor should thus be permitted temporarily—and then forbidden again.

The last sentence obviously sounds like an attempt to placate the orthodox; the key to the quotation, however, is the meaning behind the term "sufficient living," which no longer, as we shall have an occasion to see later, refers to a simple, frugal, and unostentatious life.

In the report of the general assembly for 1956, hired labor appears on the list of "unsolved problems"; it had been decided "in principle to abolish hired labor again, but to do this without causing serious harm to the farms employing them." By 1957 the tide seems to have turned, and suggestions for major changes begin to appear in the bulletin. One of the younger members proposes a complete abandonment of the principle of family labor and also advocates the employment of hired experts for the major branches of Nahalal agriculture, to achieve greater specialization and efficiency. Although these articles bring sharp reactions from orthodox members, the conclusion is clear: abolition of the principle of family labor is advocated quite explicitly and not in an apologetic whisper.

Hired labor, in fact, becomes institutionalized, and the general assembly recognizes this fact by setting formal restrictions on it: "A farm that has more than one pair of hands may not use more than two days of hired labor per week. A farm that has only one pair of hands may use four days of hired labor per week. Special cases will be discussed by the committee." [12] The flavor of the new situation is, however, expressed much better by a letter in Bulletin No. 70 (November, 1957) in which attention is called to a note published by the committee itself: "All members are requested not to pay laborers and housemaids in cash but by a note to the committee." The shame, the orthodox settler points out, is triple: "Not only is there hired labor, and not only is it officially recognized by the committee and assisted by it, but reference is actually made to maids!"

Changes in the pattern of mixed farming, like those in the principle of family labor, were a response to objective processes and were spearheaded by the young generation. Here, however, the conflict was essentially between different interpretations of the moshav settler role, both inherent in the original pattern—the image of the peasant and the image of the modern farmer. The family farm institutionalized, as was shown, the drive to achievement and intensification, and this drive was bound to mean a growing emphasis on rationality, efficiency, and specialization on the one hand, and on a progressive market orientation and dependence on the other. Nahalal had for a long time been steadily progressing towards commercialization, the very success of its family farms providing a growing proportion of marketed goods. Its market, however, had for the most part been a local settlers' market, reinforced by the wartime necessity for food autarchy; the viability of the mixed-farm could thus be maintained.

In the fifties, the economic situation changed, bringing surpluses in many basic branches, growing competition, and a general export emphasis; conditions clearly favored concentration on fewer specialized crops rather than on a balanced variety. This brought the situation to a head, sharpening the awareness of the contradiction, and differentiating between those who would reverse the process and those who wanted to adjust to it. This is again reflected in the village bulletin (1954): "An end must be put to this farming system which produces mainly for high income. We are land-tillers, not merchants. We must not be dependent on exporting our products and on buying our needs; our farms have turned into little factories."

By contrast, there are suggestions of branch limitation and of regularizing the process by which many had already hit on a profitable branch and developed it. The points commonly made were that it is difficult to be successful in all branches, and that the market is differential; inevitably, a farmer must concentrate on one or two branches only:

> We must develop flexibility in our economic ideology, and although this ideology originally demanded complete self-supply, it is no longer possible to maintain it in reality: we must decide that while the majority of needs will be supplied from the inside, some will be imported from the outside. On the other hand, much attention and scope should be given to the more promising branches. The milk farm should be given a wide scope; it should be treated professionally, and not on an amateur basis.

The quotation, reflecting the general tenor of this group, thus advocates both specialization and professionalization; it does not, however, propose any reversal of the basic family farmyard image or its transformation into a business. Unlike the case of hired labor, no formal solution has yet been laid down by the assembly, but there is a large agreement as to the permissible range of the process:

1. The village as a whole must not become a monoculture, and will thus maintain basic security and variety on the over-all level.

2. Farms can specialize, but on the condition that secondary branches are maintained.

3. The specialized branches must produce total volume sufficient to enable the village cooperative to provide suitable services to them.

As we shall see further on, another mechanism of adjustment is the establishment of partnerships, or of producers' subcooperatives, where variety can be combined with specialization.

The nature of the farm unit—as regards tenure and management and cultivation—in some respects presented a basic contradiction. It will be recalled that the twenty-five-acre unit allocated originally to each settler was double the norm considered appropriate for the family farm. This fact, characteristic of Nahalal specifically, had important future implications. It greatly increased the margin of intensification, as well as the scope for growth and entrepreneurship. While the ceiling was, as we have seen, rigidly fixed, it was much higher than in other places. These greater possibilities for development by their very nature clashed with the smallholding farm image, especially since they accelerated the introduction of hired labor and mixed farming, and fostered the economic polarization inherent in individual differences. This process had been anticipated and provision made for it at Nahalal's inception: any land in excess of family farming proper would be returned to the moshav. The idea of stripping extra fat was indeed brought up in the early forties, in the above-mentioned context of immigrant absorption. In 1941, S. Dayan writes that Nahalal and the moshav in general might be a solution to the problem of the refugees who would come after the war from Europe, and that every settler would thus have to content himself with a smaller farm unit.[13]

This solution was very much at the back of people's minds. Of course, many successful farmers were understandably unwilling to be held to this commitment and to give up land on which they had toiled for years; the question was a potential focus for a fundamental internal division. The situation, however, was resolved in the context of another problem—the second generation. As mentioned earlier, prevention of fragmentation was a cardinal principle of the moshav, and integrity of inheritance had been one of the few regulations formally incorporated in the tenure agreement. This, though, was later seen as likely to create strains and competition within the family and to denude the village of most of its young people. To prevent this, measures were taken to create new farms within the village out of the existing land. A committee established in 1957 to work out this program formulated the following conclusions:

a. In order to safeguard the existing farms, a son can get an independent farm only if at least one of his brothers remains on the parents' farm.

b. The second generation in Nahalal comprises about two hundred members of some one hundred families, 75 per cent of whom are to replace their parents in the future. The remaining twenty-five families must be provided for. It is most probable, in fact, that the number of such families will be greater, especially if we add to them the sons of the professionals and artisans. It is planned, therefore, to gradually double the number of farms in Nahalal, according to the needs.

c. The settlement of the sons will be gradual and will continue until the youngest among them grow up. The settlement and redistribution of land will be carried out accordingly.

d. In order that all the sons of farmers and professionals can settle, the veteran farms contribute up to ten acres each, 2.5 acres at a time, during the whole period of settlement. This land is transferred into public hands for this purpose.

e. A son who settles independently gets five acres from the public land and a similar area from his parents' farm. (Obviously the area which is given by the parents to their son is deducted from the amounts of land the farm contributes for the public land.)

This solution alleviated the tensions of rampant development and kept the situation within institutional bounds. At the same time, it also broke down the demographic stagnation implicit in the closed nature of the moshav system.

However, the redivision of land was bound to be harmful from the rational-economic point of view; it decreased the size of the unit of cultivation, particularly so in light of the stipulation that on each farm, specialization had to be balanced by secondary branches. But here too an innovation attempted in one area was relevant to another. It will be recalled that in the ecological layout of Nahalal, all farms had a centrally located yard and several scattered plots, care having been taken to create functional units according to crop and quality of land, rather than formal tenure. In 1951 over fifty members decided to activate the cooperative potential inherent in the land parcellation system and to create a producers' subcooperative. The recognition accorded by the assembly to this experiment was expressed as follows in 1952: "Since fifty-five farms joined for cooperative cultivation of their fields, it was decided to divide the land again, so as to enable those who cultivate their fields cooperatively to have the fields as close to each other as possible

. . . this division is temporary and constitutes an experiment in cooperative cultivation." [14]

In 1953 the farmers who participated in the experiment judged it satisfactory from all points of view, and particularly successful in resolving the problems of scattered farms and mutual aid; accordingly, they decided to buy cooperative tools and machines. By 1954, though, tension arose because of considerable differences in the contributions made by the various partners in the enterprise. A process of selection, weeding out from the experiment those who "lagged behind and were a burden on their colleagues," was undertaken. The cooperative was thus tightened up and made more efficient. By the same process, internal differentiation naturally started, with a division of labor in management of various branches. The subcooperative thus pointed to a way of solving, through producers' partnerships, the problem of specialization as regards both the viability of size and professionalization of management, again within the basic institutional structure.

So much for overproduction. Nahalal has also witnessed, although to a very small extent, the beginnings of the opposite process: withdrawal from agriculture. By this we do not mean cases of legitimate hardship, of unsuccessful farms, and of temporary decreases in the cultivated area due to approved public duties. Reference is rather made to regular outside work at the expense of farming. In practice, this was due primarily to extended party or governmental missions, usually abroad, in which members overstayed the period approved by the village and did not return upon being requested. The underlying cause was here essentially the acceptance of supralocal responsibility, implied in the moshav idea itself and discussed in relation to hired labor; it in no way implied the abandonment of the farm and of the moshav way of life. It was nevertheless considered to open the door to the most dreaded deviance of all, one common in less veteran villages: seeing in the moshav a place of residence only and letting the land either lie fallow or be put to unsanctioned uses. It is indeed in terms of this potential threat—not controllable, it will be recalled, by the formal tenure agreement—that much of the corporate problems and developments discussed later must be understood.

As we saw in our analysis of the basic ideal of the moshav, emphasis was laid not on mechanical equality, as in the collective, but rather on the equality of life chances. That is to say, recognition was given to differences in individual enterprise and success; but these differences were not meant to result in polarization. Special provisions for helping the distressed and curbing the acquisitiveness of the successful were provided and later on buttressed with progressive taxation. The main safeguard lay in the limitations of the smallholding farm itself. By the same token, however, the dramatic changes that we observed in the nature of this farm were of necessity reflected in the social sphere, with the iron law of economic growth, of the rich getting richer and the poor getting poorer. Although this process was controlled by the redivision program, the above mentioned sub-cooperative compensated, as it were, for the structural limitations; and no solution has been found to prevent the crystallization of a stratum of "too successful" farmers.

The problem of polarization is intimately related to changes in the simple way of life, once a mainstay of the moshav. This principle had initially been inevitable economically, because of the absolute priority given to productive investment over consumption and standard of living. S. Dayan thus tells of the self-abnegation that existed in Nahalal in its beginnings: "In the very first stage, when the question of dwelling came up and the means were scarce, it was decided that the farm should come first . . . first stables were built, people could live in tents . . . everybody agreed that the basis of our economic policy must be the rule to restrict expenditure as much as possible and be content with what the farm produces." [15]

However, when the first years passed and the economic problems were solved, when Nahalal became relatively prosperous, moderation was still emphasized because of the desire to preserve the essential character of the place: "The farm exists for the farmer, not vice versa. We must not become enslaved to our property." [16]

This opposition to becoming too rich, the opposition to trade and the cult of property in general, indeed accompanied Nahalal during all the years of its existence. Thus, in 1959 S. Dayan still says, "We must limit ourselves not because of a shortage of means, but because we must avoid imitation—we must create an independent, individual style of life." [17]

This fight against extravagant consumption and aggrandizement is thus accompanied by a wider aspiration, namely the wish to create a new kind of man in the moshav with a culture and a style of life different from the other forms of settlement in the country. The moshav thus defended itself, up till now successfully, against any branching off into industrialization, which, in the moshav ideology, represented the supreme evil of urbanization. In 1957 we encounter in the bulletin a severe criticism of a farmer who started the cottage industry of pickling olives: "The rural characteristics, the pure rural way of life free of any tinge of urbanization, are typical of the moshav, and Nahalal has always been the model for all the other moshavim." [18]

In other respects, however, the position had become untenable, and consumption buying of all modern luxury items, from villas and private cars downwards, had become widespread, especially among the second-generation farmers. The complaint voiced in this respect in the above-mentioned bulletin reflects the situation: "Whereas the parents were content with modest dwellings and simple dress, and dispensed with luxury, the second generation imitates the city culture and is not content with the simple way of life that characterized their elders."

Paradoxically, this trend was to some extent encouraged, even made inevitable, by the very same provisions that were calculated to prevent it: the ban on outside property and business and the ban on development of nonagricultural enterprise; in consequence, the profits from fully intensified farms, often with a gross yearly income of over £P 100,000 (about $33,000), naturally flowed into consumption spending. The same trend is also evident on the community level, with the recent construction of a luxurious community hall, the provision of individual transport for public functionaries, and the "extravagant spending on refreshment and entertainment during public celebrations."

These social changes did not undermine the essential solidarity of the community and did not affect basic commitment to Nahalal and to the moshav. They did, though, shift its basis a little from *Gemeinschaft* to *Gesellschaft*. This process is further reflected in the village organization.

The organizational structure established upon settlement and during the following years has maintained itself in its basic form. However, significant differences have developed in the functioning of the various institutions and in the relations between them, in line with the general social processes observed. Perhaps the most interesting change is the emergence of the cooperative committee as the real focus of activity in the village, as against the earlier prominence of the social-municipal council. This primacy of the economic framework reflects the "routinization" and the "secularization" of the original social drive and represents the central place now occupied by productive, financial, and marketing problems and initiative. It also represents, however, the cold reality of government. It will be recalled that the moshav has never incorporated municipally, its complex *general* administration being ultimately based on voluntary acceptance. When basic identification persists, this is of no actual import; but with the lessening of the complete commitment, the power of the cooperative—formally registered, commanding resources, and capable of controlling dues and payments through the management of individual marketing accounts—becomes crucial.

Of course, in Nahalal this function of social watchdog and tax collector was as yet latent rather than actual. However, the process of strengthening the cooperative was much more than an expression of economic enterprise and a safeguard of municipal and social authority. It was, as will have been gathered from the preceding pages, a corollary of the inroads made upon the fundamental agrarian system of the moshav—the family farm; it reflected the fear of, and the defense against, the weakening of the basic commitment to moshav values and way of life. Concretely, the trend has thus been crystallizing towards compulsory membership and towards the identification of the community with its cooperative society, which would prevent the members from exempting themselves from public control. From the point of view of the cooperative itself, such a rigid pattern, if adopted, would of course restrict its structural flexibility and prevent the emergence of possible alternative forms; for instance, intra-village cooperatives among similar specialized branches. Much more basically, however, it might signify the transition from a voluntary to a compulsory basis—more properly, to a legal contract, binding for the duration of membership.

The story of Nahalal, like that of Petach Tikva and of Ein Harod, is essentially a story of success. It is a story of pioneering and of achievement, and of personal and national fulfilment. There could be no more eloquent tribute to it than that it has served as a guiding image both to scores of direct followers, as well as to over 270 immigrant villages settled after independence.

The most striking characteristic of Nahalal has been its ability to absorb and sustain change and to tolerate internal differences. The history told is thus essentially a history of conflict between the "peasant" and the "modern farmer." This conflict—now resolving itself, apparently, in favor of the latter—was accompanied by agonizing soul-searching and tensions; but this did not undermine the essential solidarity of the community, did not affect basic commitment to Nahalal and to the moshav, and did not impair the viability of the institutional structure. Considering the nature of this structure—detailed, rigid, and highly interdependent—its ability to produce modifications was indeed remarkable. For this very reason, the story of Nahalal is one of adjustment rather than of innovation, of grudging acceptance of change, and not of purposive social experimentation. In this respect, at least, of all the forms analyzed, the moshav is the most conservative.

Postscript

After our analysis, based on data carried into the late fifties had been concluded, the fundamental problem of the future path of the moshav crystallized. As was made clear above, in Nahalal itself the threat to the moshav way of life was potential rather than actual: changes were carried out in reference to the basic institutional structure. While the young chafed under the restrictions imposed upon economic activity, essential commitment was maintained; there was no leaving and few cases of actual rejection of the voluntary social control. Not so, however, in the movement as a whole, and especially in the ideologically uncommitted new villages, established from among the mass-immigration after the War of Independence. As we shall see later, the movement hierarchy felt that these deviations endangered the very soul of the moshav; consequently, its voluntary and informal mechanisms of social control were now insufficient. To meet the situation a moshav article was proposed for inclusion in the newly prepared Bill on Cooperative Societies that would legally compel the settlers to become members of the cooperative and would give the community strict control over them. The tabled regulations concerned: membership of the moshav; possession of property in the moshav by one who is not a member; and expulsion of a member of the moshav.

The regulations at once aroused considerable opposition and are, in fact, still being debated. The items that caused the most violent argument were those concerned with the cessation of membership of the moshav. The proposed law demands expulsion in the following circumstances:

1. The member has a farm but does not work it or is a member of the family of the farmer, under section (3) of article 171, and does not cultivate it, and this without permission of the general assembly of the moshav as set out in article 172 (a), or by a

breach of the conditions of permission granted by the assembly, its decision being final;

2. The member regularly neglects his farm without sufficient cause;
3. The member deliberately damages property or equipment received to facilitate settlement, which belong to the settlement institution or the moshav;
4. The member disturbs the peace by violence or the making of threats, and has been found guilty by an authorized court;
5. The member permanently goes to live outside the area of the moshav, without permission of the general assembly of the moshav, as under article 172 (a), or by a breach of any of the conditions of permission granted by the assembly, its decision being final;
6. The member fails to fulfil those obligations laid down in the regulations of the moshav defined as compulsory for all the members, the breach of which is accepted as grounds for the cessation of membership of the moshav.

A Nahalal minority initiated and led this opposition, its most vocal member being—ironically, but perhaps understandably—the son of the founder of the moshav. To them, the problem at stake was that of individual freedom being subordinate to an external law. They saw the greatness of the moshav movement in the voluntary submission of its members to rules and restrictions; enforcing the law, they believed, defeated its own ends: "Enforcement turns a principle into a nightmare; when a person *must* act in a certain way, there is no virtue in his action. The danger of ceasing to exist as a moshav is greater than this exaggerated adherence to the principles, which would even resort to pressure and compulsion, and which relies on the fear of sanctions. In ideological problems none has moral authority to exert physical pressure." [19]

The rebels thus agreed upon adherence to moshav principles as such and maintained that their argument was not concerned with the contents of these principles, but with the way they were maintained. The rebellion led, nevertheless, to a period of social ostracism, to broken friendships, and, it is said, to economic sanctions.

By the time this is being written, the severe social crisis in Nahalal seems to have spent itself; the community, while agreeing to differ, is again functioning as a whole. Not so, however, the large problem itself, in respect to which the moshav is still at its crossroads.

This question, though, belongs to the movement at large and must abide its turn.

PART II

ORGANIZATIONAL DEVELOPMENT

5

Growth of Colonies

This chapter deals with the organizational frameworks developed by farmers in the colonies. The analysis avoids chronological details and emphasizes the salient features that characterize the colonists' organizational activities and set them apart from the collective and cooperative movements examined later. Particular attention is directed to the essentially instrumental basis of farmers' organizations, to their relative lack of political and ideological objectives and organs, and to the dominance of local, regional, and sectional interests in them, with their consequent inability to merge into countrywide bodies and wield influence upon the central institutions. Some of the factors in terms of which these features may be explained are traced, chiefly, the lack of ideological motivation and missionary zeal of the farmer, his traditionalist and narrow horizons, and the restrictive conditions—philanthropic and political—under which his organizational efforts were first made.

During the period discussed (1880–1948) three regional organizations of moshavot were at different times active. Two of them, the Judaean and Samarian, and the Lower Galilean, were permanent in character, while the association of the Sharon colonies was much more sporadic. (See Map 5). Our discussion focuses chiefly on the first of these, the Judaean-Samarian (henceforth designated Judaean), which has been politically and economically the most important and which is also directly related to Petach Tikva. At the same time, a brief attempt is made to trace the development of the other frameworks, too, so as to explain the causes of their differential organization on separate local basis.

The chapter falls into four parts: the period of 1880–1900, distinguished by complete atomization; the period of 1900–1920, featuring the colonies' first efforts at organization; the years 1920–1930, in which the organizational frameworks are set up on a regular basis; and, finally, the phase of intensified activity between 1930 and 1945.

The First Period—Tutelage (1880–1900)

This period (in which mainly Judaean colonies existed) was characterized by the absence of any formal supralocal framework maintained by the colonies themselves, either vis-à-vis each other or vis-à-vis the external authorities. The central organization was in each case the village council, which dealt exclusively with internal community matters: cultural activities, land registry books and cadastral surveys, legal proceedings, and guard duties. True, the colonies were also in political contact with the Turkish government, but these were personal contacts of local notables with officials, in which no community as such, and certainly not the moshavot as a corporate body, had any standing. This type of organization, which possessed a measure of internal autonomy (chiefly cultural) but no representative political status, reflected indeed the practice of the Turkish Empire in respect to its subjects, and especially the minorities. Indeed, the political structure of the Ottoman Empire was not based on the consolidation and articulation of grass-root political power, and contact between the authorities and various local and ethnic frameworks was provided by individuals or notables, who personally intervened on behalf of their groups.

The activity of a personage like Baron Rothschild, a powerful and influential man, was thus better suited to the Turkish system than the activity of community representatives or organizations. Therefore, although each colony organized autonomous political bodies that strove to fulfil the internal needs of the community, any organizational framework above the colony level could not gain proper status vis-à-vis the Turkish authorities. No economic activity or economic cooperation between the colonies developed either. Such patterns usually arise in answer to economic challenges, but these did not exist here: the sale of agricultural produce (wine) was controlled and safeguarded by the Baron's officials, who also established and managed the wine cellars for the processing of the grapes, so the farmers never experienced the economic difficulties involved in production and marketing.

Thus, the period of Rothschild's patronage over the Judaean colonies was not only responsible for their narrow monocultural basis and for the sapping of their farmers' initiative and independence, as in Petach Tikva; the patronage, tailored to the Turkish political system, led to organizational and political apathy.

The Second Period—First Steps (1900–1920)

For most of this period, Palestine was still a part of the Turkish Empire. Since the over-all political framework remained as it was before, no change occurred in the political organization of the colonies. Activity began, however, in the sphere of economic organization, and in this period the first attempts to create economic frameworks transcending the limits of the individual colony were made, only some of them proving permanent.

The first organization established in the Judaean colonies was the citrus-growers' cooperative, Pardess (Orchard). It is significant that the initial steps were taken in a branch in which the Baron's administration had little interest, and which, unlike the vineyards and the grape industry, it had not dominated. Thus, while the first citrus groves had been planted by individuals in the eighties, they enjoyed no economic support, professional guidance, or help in marketing as the vineyards did. In 1900 these plantations were concentrated mainly in the Judaean district and stretched over an area of five hundred acres, of which the Petach Tikva farmers owned about half.

J.C.A. was a member of Pardess from the outset, but it was

managed, then and later, along democratic principles, with an elected board of directors responsible to the general assembly of the cooperative and meeting once a year.

The first objective of Pardess was the cooperative marketing of the fruit. The aim was to dispense with the services of Arab agents and their primitive methods and limited traditional markets, which, for the Jewish citrus grower, were proving unprofitable and hardly covering production expenses.

The following innovations may be noted in the sphere of marketing: the opening of marketing agencies in the big European centers, for purposes of market research and extension of marketing possibilities; the acquisition of a shipping agency to intensify control over the quality of fruit-transportation; the establishment in Jaffa of a warehouse for the storage of fruit before loading, so as to become independent of Arab warehouses; the renting of a wood from J.C.A. and establishment of a crate factory, in order to minimize the costly import of wood; and research designed to improve packing methods, in order to ensure the arrival of the fruit in good condition to markets abroad.

For the fruit-bearing trees themselves, the cooperative brought experts from abroad and carried out research in order to improve methods of cultivation and pest control. The main financial resources of the cooperative—until payment for the fruit arrived from abroad—came from two sources: the Anglo-Palestine Bank and J.C.A. The cooperative's relationships with these two institutions were purely economic, the loans being given under the usual commercial conditions and guarantees.

Pardess thus was an innovation in two ways: it was a cooperative, and it was an independent body, profit-oriented and based on own capital. In retrospect, the creation of the first cooperative of the moshavot was perhaps the most important single agricultural (as opposed to settlement) factor during the whole period under consideration.

The second attempt at independent economic cooperation was made in 1906, this time in the then dominant, grape-growing branch, through the establishment of the Agudat Hacormim, an association of grape-growers and wine cellar owners.[1] It will be recalled that the wine branch had been dominated by the Baron. Therefore, the first independent organization in it was created after his departure and the transfer of the management of the colonies to J.C.A. As was seen, J.C.A. employed healthier economic policies than the Baron's officials, giving credit instead of philanthropic support and forcing the farmers to make autonomous decisions involving greater risks. The new cooperative association grew out of these risks and provided some security against them.

This association, unlike the Pardess cooperative, began its activities on the basis of previously established frameworks that were not its own creation: the wine industry complex existed already and had reached a high standard owing to Rothschild's initiative; Palestinian wine already possessed secure markets in the U.S.A., in central and western Europe, and in the Middle East. However, the element common to the two organizations under discussion was cooperation designed to promote a specific economic interest, and neither their professed aims nor their activities transcended this limited sphere.

A broader field of activity emerged in the Judaean district later in the period. The first political association of the farmers was the Federation of Judaean Colonies, set up in Judaea in 1913. This was a federative organization, which first united the following colonies: Beer Tuvia, Gedera, Ness Ziona, Beer Yaacov, Rishon Le Zion, Petach Tikva, and Rehovot. In its organization and the extent of its functions, this federation differed both from the farmers' organization in lower Galilee and from the previous Judaean organizations such as Agudat Hacormim or Pardess; its basic unit was the colony and not the individual farmer, while its aims transcended the narrow goals characteristic of the organizations hitherto dealt with, as can be seen from some relevant selections of its statutes.[2]

The aims of the organization:

a. To hold and protect the political rights which the government laws grant to the colonies. For this purpose we have chosen one representative for all the colonies, and he will be our spokesman before the government in all matters affecting the colonies. No single colony is entitled to negotiate with the authorities about general matters unless the organization's council and its general representative agree to it. . . .

c. To try to create good relations between us and our neighbors [the Arabs] and to disprove the lies told about us.[3]

d. To strengthen the colonies' economic basis by means of various syndicates on cooperative lines, improvement of livestock, road repair, etc. . . .

We hope that our functions will grow and multiply as our work progresses and will embrace the whole of our spiritual, economic, and political life.

Two main motives seem to have been responsible for establishing the federation: the tense political atmosphere before the outbreak of the World War, which the farmers felt acutely, owing to ever new directives issued by the central Turkish authorities (additional taxation, mobilization orders), and which created the need to put up a united front of resistance; and more general economic problems, transcending the specific boundaries which Agudat Hacormim or Pardess set for themselves, and arising with the gradual but definite severance from the J.C.A.'s protection.

The main financial resources at the organization's disposal were internal, and therefore we find in the statute: "The organization's fund accrues from an annual payment of ½ per cent of gross income [of each colony] from plantations and ¼ per cent from field crops. This payment is effected half yearly." Further on we shall see that this paragraph constituted a serious obstacle in the federation's development, since it was not adhered to by the colonies. But it is crucial in another way as well: it points to an independent course of action that the farmers had to take because they had no access to the financial resources concentrated in the Zionist Organization's Treasury.

At the time of the establishment of the Palestine Office in 1908, it did not yet have any fixed agrarian or settlement policy;

but the moshavot had made no effort to influence the program as it was being formulated or taken an organized part in it. Not that the pattern of the private colony could become, under the existing conditions, the main public settlement tool and the primary receptacle of Zionist funds. However, *no* attempt at all was made to push its claims, and certainly none to modify it, so as to lead in the establishment of new forms, which could have enjoyed the contemporary preference for family-based rather than revolutionary villages. Of course, this was almost inevitable in the light of the colonies' background and history, and their lack of any missionary ideology and drive. Indeed—as we saw in the case study of Petach Tikva—they were economic entrepreneurs, but by no means social innovators, and still lived largely in terms of the closed and isolated world of the traditional Jewish village in the Diaspora. Moreover, the potential chance came their way when they were organizationally unprepared and just starting on their independent development.

Be that as it may, the Judaean moshavot—centrally located, and growing populous and rich—became marginal vis-à-vis the main efforts and institutions of the community, and sometimes even opposed to them. This relationship was, as we shall see, to persist for many years to come.

Indeed, the political activity of the Judaean colonies solely aimed at easing the burden the Turkish government imposed; it was carried on through informal channels, usually through the intermediary of the J.C.A. representative in Palestine, who had extensive personal contacts with senior Turkish officials. Meetings between the federation's representatives and the informal Jewish élite—the permanent committee, whose members were townsmen for the most part—were not infrequent, but they provided a channel of communication and integration rather than a framework for the presentation of demands. On the other hand, contacts with the labor power centers produced a clash of values and interests. We have seen this in Petach Tikva, which clearly exemplified the struggle between the militant socialist and the man of property, who, while identifying himself generally with the national revival and contributing to it greatly by his very entrepreneurship and productivity, viewed the worker as a danger to his all-important economic interest. This collision existed indeed throughout all the Judaean moshavot and became most severe in the spheres of labor and defense.

As mentioned in the case study, the farmers preferred the cheap Arab workers from outside the colony to the Jewish ones, whose wages were much higher. It should be remembered, however, that the economic considerations, though important, were not the only ones; the farmers were also afraid of the impact of the workers and of their way of life, and anxious not to lose their exclusive local power positions. The farmers thus organized politically not in an expansionist drive, but chiefly in response to external pressures. Two challenges in particular motivated them: the political situation and the need to unite to meet war conditions; and the growing organization and power of the Jewish workers in the colonies.

This passive or defensive nature of the farmers' organization, without any militant ideological drive, was reflected in internal weakness and relative lack of political influence. In 1920, when the first assembly of representatives was elected, the farmers' federation put up its own list and sent sixteen delegates out of a total of 314 places [4]; some of its prominent leaders were included in the board and in the plenum of the executive committee of the assembly. But this fact had no great political significance, since, as was mentioned in the introduction, the assembly only dealt with such matters as health, welfare, and culture and had no political or economic competence. Even in these limited spheres, it is doubtful whether the farmers' federation was capable of effectively furthering its demands, not only owing to its small representation in the assembly, but also because by the end of the war its weakness was clearly manifest.

In the meetings of its center held after the war, its statutes were discussed and its economic and political objectives reaffirmed, but "there was great enthusiasm, great hopes, but the good intentions were not realized. The resolutions were not carried out. . . . Many members were no longer interested in the federation. After a few flickers of life, the organization ceased to exist for many years. . . ." [5] As one of its leaders put it: "It was the farmers' indifference that nullified the attempts at cooperation for a long time. Only the difficult conditions prevailing just before and during the war shook them out of their indifference —a phenomenon similar to that noted in the organizational attempts in lower Galilee—but when these passed, the apathy returned." [6]

The Judaean and Samarian colonies began to prosper economically owing to the development of two promising agricultural branches (wine and citrus) and thus grew in economic importance, finally achieving economic independence with the severance of their ties with J.C.A. at the beginning of the twenties. The economic conditions of the colonies in lower Galilee, on the contrary, were bad because of the distance from urban centers, unfavorable climatic and soil conditions, and shortage of water, all of which delayed for a long time the development of their mixed-farm economy. In consequence, the patterns of organization that they developed were unable to protect economic interests on a countrywide scale or to serve as a basis for political demands, as was the case in Judaea. The more so since, as we shall see in the following chapters, mixed farming became the realm of the workers' settlements (kibbutzim and moshavim), which they dominated in respect to specialization, professionalization and experimentation, volume of production, and marketing.

Indeed, in so far as the Galilee colonies had any general economic interests at all, they meant, at least initially, no more than survival. Economic conditions bordering on poverty, which existed for a long time, and the constant need to defend the homesteads from Arab depredations in common with the collectives of the district also prevented the creation of any strong class-consciousness or, more properly, as happened in Judaea, the crystallization of private employers' interests leading to political struggle.

The Galilee colonies were in any case prevented from participating in such political struggle, even in the narrow sense of competition for scarce resources, because they were largely marginal to the centers and channels of political power. They were economically and administratively dependent on J.C.A., an organization both formally and in practice apart from the Zionist movement, and thus external to the Jewish community in Palestine. Up to the forties, J.C.A. remained the dominant center to which most of the colonies' demands were presented, except perhaps in the sphere of security, which enjoyed special attention on the part of the Yishuv's political institutions because of the Galilee district's great vulnerability.

Thus, the area's slight economic importance, combined with prolonged dependence on J.C.A., reduced to a minimum the colonies' participation in the Yishuv's institutions, which began to take shape after the First World War, and reduced situations of contact between the farmers' representatives and these institutions. On the other hand, different economic interests, which prevented the development of the capitalist image of the landlord, hindered any organizational cooperation with the Judaean colonies, which possessed a clearly defined Right-wing political orientation.

The first attempts at organization appeared in the lower Galilee district in the first decade of the present century, when security conditions deteriorated following the weakening of the central Turkish government. The colonies were constantly subjected to Arab raids, robberies, and murders, and this led to collaboration between them. However, this collaboration was for the most part sporadic, aimed at defense and guard duties, and did not evolve definite patterns. In fact, the farmers relied mostly on the help of the Hashomer organization created by the workers' sector (described later on). Conflicts tinged with class consciousness, which accompanied the Hashomer's activities in the Judaean colonies, were absent here.

An attempt to improve the backward economic situation of the colonies in the Galilee by means of cooperation was made in

1912, with the establishment of the Agricultural Organization in Galilee, a cooperative institution designed to foster cooperation in the spheres of supply, marketing, cattle insurance, and improvement of cultivation methods. However, this organization was no sooner set up than it collapsed, owing to the farmers' lack of interest. This indifference was due to conservatism: "In the nature of things, the organization was not able to carry out the aims it had set itself and the Galilean settler, whose mentality was similar to that of a *fellah* [Arab traditional peasant], was not interested in public affairs unless and until a bullet was flying from ambush or a greater calamity alarmed and unified." [7]

In 1914, when the security situation continued to deteriorate, a new attempt at political organization, including the workers' sector, was made: "This assembly confirmed that the organization should be a permanent body representing the colonies before the government and charged with defending lives and property in the colonies. . . ." [8] Membership in the new organization was individual, and each farmer was obliged to pay a membership fee of ten francs a year.[9] The assembly, which took place in Yavniel, also fixed the number of fee-paying members from each colony who could vote.[10]

This organization had wider aims than those of former Galilee associations. In addition to protection proper, it financed the exemption of farmers from Turkish army service and negotiations for reduction of the heavy taxation imposed on the colonies by the Turkish government. Its political pressure was also directed to the Turkish government, but in contrast to Judaean associations, it had a nonsectoral character, with labor organizations as members, a phenomenon that can be understood by the lack of the landlord image among the Galilee farmers; the main resources of the organization were own money and loans from J.C.A.

This common political activity lasted, however, only as long as the emergency created by the war; it then became completely marginal. Moreover, the representation of Galilee farmers in the Jewish community's new autonomous political institutions, established at the beginning of the mandatory civil administration in Palestine, was only symbolical. The reason for this was that the farmers were not affiliated to any important political party. (The Progressive List, which sent one of the farmers' main leaders to the assembly, was a transient phenomenon, devoid of national significance.) While the organization of colonies in lower Galilee changed form several times until 1923, economic activity became dominant in its permanent patterns, to which we shall return later on.

The Third Period—Consolidation (1920–1930)

The organizational relapse, brought about by the anticlimax of peace, continued for several years, after which activities were resumed, this time on a firmer, more regular basis. In Galilee, the year of 1923 thus saw the emergence of a new protective organization, following a redefinition of aims.

Some of the members of the earlier framework established in 1914, which operated during the war, now argued that the organization should become a true farmers' association; the labor elements should be excluded and the concentration of its activity transferred from political and security matters (the basis for the close cooperation between farmers and laborers during the war) to economic matters, in order to promote the farmers' immediate interests. Others felt that the composition of the organization should be left as it was during the war. This conflict of opinions went on for four years and effectively destroyed any chance of common regional activity; in 1923, however, the innovators won and the Mutual Economic Association of the Farmers in Lower Galilee, Ltd. was set up.

Although the new organization was indeed composed of farmers only, both factions stressed its apolitical and nonpartisan character. This image of the new organization is not to be wondered at; as explained above, the Galilee farmers were daily confronted with the problem of survival, pure and simple, and their economic significance for the country as a whole was slight, so they wielded no political power. Moreover, they were still dependent on J.C.A., a company formally not associated with the Zionist Organization's institutions; this reduced still further any chance of gaining a hearing in these institutions. In any case, the economically separatist and autonomous character of the organization was only formal, and in practice intersectoral cooperation continued.

In 1929 the farmers' association even passed a resolution that obliged its members to employ Jewish workers exclusively. Although this resolution reflected the security situation then prevalent (the 1929 Arab disturbances) and the economic changes that the association intended to introduce in the colonies,[11] it nevertheless showed a much less militant class interest and social image and a greater national integration than those shown by Petach Tikva and its fellow colonies.

As in the former period, the Galilee organization had no political aims, and even the presentation of a list to the second assembly of representatives (1925–1931), out of which one delegate was chosen, should be viewed as a symbolic gesture, showing a wish to integrate in the community and its institutions, and not as motivated by a desire to gain power and influence in the assembly. Indeed, from 1931 onwards, the farmers of lower Galilee were not represented in the Yishuv's political institutions. Neither had they any representation in the Jewish Agency.

In 1923 a new organization was established in the Judaean district, the Association of Jewish Farmers in Palestine.[12] There is no doubt that one of the main reasons for reviving the organization was the farmers' need for a better integrated group capable of defending its specific interests on the new political and economic scene of the Yishuv. Unification was in fact inevitable in the light of the creation and expansion of the Federation of Labor[13] with all its economic, political, and social institutions, which charged the sectoral labor struggles with intenser meaning. The motivation to organize was now, even more than before, the necessity to protect vital interests.

We suffer both spiritually and physically in spite of the fact that we farmers are the chief builders of our country. Even though for the past forty years we have been fighting and overcoming various difficulties and obstacles, our existence is not sufficiently felt. . . . We are dispersed and separated, each in his corner, with his troubles and despair. Teachers and clerks are organized; workers and craftsmen have their frameworks, and only we farmers, in spite of all attempts that went before, still remain in our initial situation. . . .[14]

Formally, the new organization did not differ in its aims from its predecessor, which had been established on the eve of the war. It too assumed two goals: economic (development of agriculture, encouragement of cooperation in the sphere of credits, marketing and supply) and political (representation of farmers before the mandate authorities and organizing against the workers' interests). These aims are clearly formulated in the new statute:

To develop agriculture in Palestine by improving the existing farms and by introducing new agricultural branches; to develop the household economy and the irrigation network.

To develop the sale of the produce in the best possible ways by arranging for partnerships and by finding markets in the country and abroad and to create agencies in the cities to handle the farmers' produce.

To offer credits to members by establishing local agricultural loan funds.

To establish an agrarian bank.

To provide members with the most modern implements, best suited to conditions in the country, as well as materials such as fertilizer, seed, and others from the original sources.

To request the government to introduce order into and to improve all the burdensome laws concerning agriculture, and to promulgate such new laws as will encourage the development of agriculture.

To negotiate with the mandatory authorities concerning the methods of taxation and the arrangement of documents relative to land ownership.[15]

In its statutes the new association declared its intention of dealing with the mandatory government directly, even though at the time these aims were formulated the Jewish community in

Palestine already possessed its own political instruments. These instruments, and especially the Jewish Agency, aimed at representation of the central political and economic interests of the Yishuv before the new authorities. The by-passing of the Jewish political institutions and the employment of direct pressure on the external center becomes quite a frequent phenomenon in the thirties and in the forties, and we shall have occasion to refer to it again.

A vital difference between the Federation of Judaean Colonies and the new association was in the form. While the Judaean institution had been a *federative* organization of colonies, with the colony council representing each colony on the central institutions of the federation, the new farmers' association was based on *individual membership* of farmers, without the intermediary of the colony council. Furthermore, the association now contacted its members not through the colony councils, but through agricultural councils, new bodies which just then began to emerge in the colony. The colony councils of necessity dealt with matters affecting all the inhabitants of the colony as a municipal unit. This population gradually became differentiated by the addition of laborers, merchants, craftsmen, and others. The fear arose that the intrusion of these elements might lead to the council's neglecting the farmers' specific interests. One of the main requests put before the mandatory government by the farmers' association was the formal recognition of the agricultural council and the grant of legal authority so that it could better defend the farmers' interests.

As regards financial resources, the new organization, like its predecessor, was dependent on internal support; but since the citrus-plantation owners were prosperous and continually gaining in national economic importance, the organization, as the citrus-growers' spokesman, could command internal taxes as well as credit on a much greater scale.

It is important to stress that although the Association of Jewish Farmers in Palestine became, as it were, a natural continuation of the Federation of Judaean Farmers, and farmers from that area took part in its establishment, the character that its founders wished to impart to it was not a local one; it was thrust upon the organization against its wish. In the 1923 statute it was said that one of its aims was to include *all* farmers in the country; in this aim, however, the organization was only partly successful. The association of lower Galilee colonies thus kept its organizational independence throughout. We have seen that their special economic interests and problems and their more national and left-of-center orientations conflicted with the aims and purposes of the Judaean organization.

Similarly, the Association of Jewish Farmers in Palestine did not attract the members of the National Farmers' League, an organization that operated after 1926 in the same district, until 1937.[16] The League was based mainly on the young colonies established in the Sharon Valley in the postwar years to ride the crest of successful citrus development, pioneered by the older Judaean moshavot. In a large sense, these colonies were the first and only truly private enterprises, being founded on private initiative, by private capital, and on private land.[17] They were also novel in their emphasis on the most profitable investment of capital, an initial motivation for settlement. The two organizations thus had identity of economic interests and seemingly similar social orientations. However, while the farmers of the old, established colonies in Judaea tended to employ Arab labor for economic reasons, the Sharon farmers had a positive attitude towards Jewish labor and supported its national claims.

The farmers of the Sharon colonies were indeed closer in their social background, essentially secular views, and Zionist ideology to the main stream of immigrants now coming to the country; they thus had a less extreme group consciousness than their Judaean colleagues, recognized the right of the Jewish worker to organize, and approved the aims of the Federation of Labor. The League existed up to 1937, when it merged with the Association of Jewish Farmers in Palestine. (It first did so in 1929 but left again in 1933 because it did not accept the Association's attitude towards the question of Jewish labor.) The League stipulated that its members serving on the board of the association would have absolute freedom of opinion in all matters of labor policy, whenever the board would meet with outside agencies, such as the Federation of Labor or the Jewish Agency.

The League's decision to give up its organizational independence stemmed from its political and economic weakness and from a feeling of dissatisfaction with the Federation of Labor, with which it identified but which rejected its advances. When there was great activity in the building trades in the cities in the thirties, laborers streamed from the Sharon colonies to the cities, where they could get a higher wage; the League's request to the Histadruth to help solve this problem did not meet with any success. Thus, the merger with the Association of Jewish Farmers in Palestine was the outcome of failures to solve manpower problems and of frustration generated when the bodies with whom the League identified ideologically failed to respond to its pleas.

Integration with the association also gave the League members better access to facilities vital for their economic development—marketing, supply, and credit; these could be had through cooperation with the old established citrus branch, which was represented in the association. However, this was by no means a political-ideological alliance, and in this respect the Judaean colonies were still standing alone.

The Fourth Period—Intensification and Strife (1930–1945)

The final period witnessed a great intensification of activities, with increasing prominence given to mobilization of resources and to direct political participation. However, before examining these developments, sustained chiefly by the Judaean association, something must be said about the Galilee organization.

It will be remembered that the most acute problem with which this organization was confronted from the very beginning was the colonies' bad economic situation. Traditional field-crop farming, lack of diversification in agricultural branches, primitive methods of cultivation and frequent shortage of water had handicapped the colonies' economic development. This even led to the abandonment of farms. The leadership of the Galilee association reached the conclusion that only intensification and mechanization of cultivation, together with a diversification of crops, could rescue the farmers from their stagnation and destitution.

This leadership asserted that such changes would also make it possible to enlarge the number of Jewish workers employed in the colonies, a development they viewed favorably. J.C.A. acceded to these demands by putting the necessary resources at the organization's disposal. The farmers, however, showed open antagonism from the beginning and, when the innovations were not immediately successful, held the leaders to be responsible for the failure. The relative lack of success of the Galilee association and the fact that the Galilee farmer has, in many respects, remained the least modernized are the result of this conservatism.[18]

In addition to intensification and organization of production, the leadership also tried to encourage cooperation in marketing and financing as a means of improving the situation of the colonies. But in spite of these efforts, the cooperative activities of the local farmers never became extensive, and most of the produce continued to be marketed through the institutions of the Federation of Labor.

In the forties, the organization increasingly shifted the direction of its demands from J.C.A. to the Jewish Agency and to the Jewish National Fund, both institutions set up by the Zionist Organization. At that time the formal contracts between J.C.A. and the local farmers were finally signed, and the farmers' organization did much to formulate the conditions of these contracts. They provided for the transfer of ownership of land and equipment to the farmers, in return for a future settlement of outstanding debts. The organization now tried to utilize the farmers' freehold to use their lands in a way that would, in its opinion, accelerate the colonies' development. Since the new methods of cultivation enabled the family to gain a decent living out of a smaller plot, and since some of the farmers no longer wanted to cultivate their farms at all, the organization's idea was to sell some of the land to the Jewish National Fund. The sale was to be conditional on the Jewish Agency's settling kibbutzim or moshavim on this land, *contiguously with the lands still held by the farmers in the colonies.* The organization thus hoped that geographical proximity to more progressive agricultural patterns would help the farmers to learn modern methods and to acquire healthier economic orientations. But the plan failed, because the workers who settled on the land tended to hold aloof from the farmers, and contacts were few.

While the Galilee farmers' association concentrated only on economic problems specific to the area, the Judaean organization engaged in manifold activities, extending to many spheres. With the worsening of the security situation because of Arab disturbances in the late twenties and the thirties, positions of power in the defense institutions of the Yishuv, and with them access to supply of arms and influence on defense policies and priorities, increased greatly in importance and became the foci of struggle. The Judaean farmers' association, for this purpose aligned closely with the urban citizens' bloc (see later), took a large part in this activity.

The farmers in the old, established colonies in Judaea had been extremely wary of the Hashomer organization, fearing its possible influence in the colonies and also differing in their conception of defense arrangements and methods. The farmers had been afraid, moreover, that a countrywide, permanent Jewish defense organization would, in determining its policies, constantly disregard purely local considerations and thus jeopardize economic interests connected with the daily contact with the neighboring Arab population. Because of this attitude, the farmers did not participate in the Yishuv's defense activities for a long time and did not take part in a new countrywide Jewish defense organization, the Hagana, initiated in the twenties by the workers' party, Achdut Haavoda,[19] to take the place of the Hashomer, which was dissolved in 1920.

In the thirties, however, when the situation deteriorated considerably, and local measures and interests could not form a basis for action, the Judaean farmers' association sent a permanent representation to the Hagana headquarters, a political body in which left-and right-wing parties were represented on a parity basis. This body was responsible for policymaking, for defense budgets, for acquisition of arms and their distribution to local units. But like the other right-wing elements in the headquarters, the farmers' bargaining position was weak in comparison with that of the left-wing parties, who had founded the organization, were much more united, and took care to assign their best men to the important professional jobs in the organization. They successfully demanded for their settlements priority in arms, budgets, and participation in policymaking.

Charged with making the Yishuv's political decisions, it was easy for the Jewish Agency to demand control of the defense institutions as well, and thus to change the balance of power in favor of the left. Also, the representatives of the farmers' association in the central defense institutions reflected primarily narrow and well-defined sectoral interests, rather than the general one. At the same time, defense was one sphere in which the association accepted the supremacy of national considerations and was ready to limit its own freedom of political action.

The colonies had from the beginning been marginal to the organized efforts and institutions of the Zionist movement. The weakness of this link resulted essentially from historical factors, chiefly the initial dependence upon external bodies and a fundamentally narrow, traditional, and nonexpansionist social horizon. As a result, it was left to the workers' movement to put at the disposal of the Zionist Organization the initiative, the manpower, the organs, and the patterns of settlement; this created the close tie between them and promoted the workers' decisive influence. The marginality of the moshavot vis-à-vis the Yishuv also meant, however, that they could not count upon and mobilize the political influence of the urban right-wing sector as a whole. In any case, this sector had nothing like the organization of the workers' movement: its main party, the General Zionists, was essentially an "assembly of notables," without any significant political apparatus. It was a relative newcomer on the scene, composed mainly of Fourth and Fifth Aliya members; and it was internally split into three distinct factions. In spite of its great economic weight, it was a rather weak minority group.[20] Even more, this group was fundamentally a Zionist one, and it consequently could not admit and support the narrow economic claims and interests, often local and sectoral, of the Judaean farmers. In most issues, such as those of Jewish labor and of security policy, the majority of the nonlabor elements consistently sided with the official policy; the moshavot could count only on a small segment, called the citizens' bloc.

The farmers' association, and the interests it represented, thus remained largely outside the Yishuv's political framework. The association withdrew from the election campaigns to the third (1931–45) and the fourth (1945–1948) assembly of representatives (Havaad Haleumi). The boycott was a protest against this institution's lack of response to the association's repeated demands for political recognition of its *economic* importance. The point in dispute was the election system, which was based on universal suffrage and proportional representation, the accepted mainstays of democracy. Under these principles, however, the association obviously could not gain a representation appropriate to the economic interests and achievement it stood for. It felt especially slighted when the more numerous, but much less productive, traditional ethnic communities, whose public importance it considered small, wielded greater influence and were given greater consideration. The demand to by-pass formal democratic principles and admit corporate representation instead was violently rejected by the left, and the farmers in effect withdrew.[21]

Since the withdrawal was only partial, and since the assembly of representatives was not a major center of power, the boycott can be viewed as a breach of the solidarity of the Yishuv much more than as an effective political act. Still, boycott was important because the farmers' association, instead of being represented by the assembly and the Jewish Agency, increased its independent contacts with the mandatory authorities and thus by-passed the autonomous institutions of the community. This was politically very significant, because while the Jewish Agency was by virtue of the mandate the formal political representative of the Jewish community before the government of Palestine, its authority was based on voluntary support—which was now breached.

The demands presented to the mandatory authorities by the association centered on three main subjects: export duties and regular taxes, municipal problems, and immigration policy. As the representative of the citrus-growers, who were dependent on exports, the association demanded that steps should be taken to free the Palestinian exports from protective tariffs abroad, and that Palestinian citrus-exporters should enjoy the same preferential status in Britain as the colonies of the British Empire did. These demands, however, were not granted. In the municipal sphere, the association strove to protect the farmers' interests in a period when the mandatory authorities were preparing legislation in this domain. The association put forward two demands, both of which were designed to limit the influence of "foreign" elements in the colony: the limitation of the voting rights in colony council elections to permanent inhabitants; and the establishment of another municipal body—village committees with far-reaching legal authority and composed of farmers only. These would take over the functions of the existing agricultural councils, which had no legal status. This last demand in particular was motivated by the desire to remove matters affecting farmers from the jurisdiction of the general municipal authority, in which nonfarmers were represented, and to entrust such matters to a body composed exclusively of farmers. After protracted negotiations and after a considerable opposition on the part of the Federation of Labor, the demands in the municipal sphere were granted.

Finally, the association demanded from the immigration department of the mandatory government a quantity of certificates in addition to the quota allocated to the Jewish Agency,[22] to enable it to recruit and regulate the manpower potential of the colony workers. This was a further attempt to end the prolonged conflicts around the problem of Jewish labor, which created tension inside the colonies on the one hand, and between the association and the Federation of Labor on the other. The association argued that it did not object to Jewish labor on principle, but it had to hire Arab labor because the conditions which the Jewish workers demanded, with the Federation of Labor's backing, were such as the farmers, and especially the citrus-growers, could not afford. In demanding supervisory rights over recruitment of the workers sent to the colonies, the association wished to free itself from its considerable dependence on the Federation of Labor, which remained the chief source of manpower supply in various economic branches, and, at the same time, to ensure that this manpower would not be under the ideological and practical control of the Federation of Labor. Therefore, in addition to demanding a special quota of certificates, the association requested that the immigrants coming under this quota should be from Middle Eastern countries or relatives of the farmers in the colonies, both groups relatively untouched by left-wing ideology and not tied to the main channels of recruitment of the Jewish Agency, the Federation of Labor, and the left-wing parties or their youth movements abroad.

In the period under consideration, cooperation in the citrus branch developed conspicuously. Besides Pardess, the largest cooperative in the branch, several other cooperatives were established in the thirties, especially by relative newcomers to the citrus branch. The fall of prices on international markets and the increase in the number of plantations in Palestine endangered the existence of the whole branch. Thus, the various cooperatives were forced to unite in order to liquidate competition, to present a unified front on world markets, and to minimize the expenses involved in the preparation of the fruit before shipment abroad.

Table 20. Citrus export of the colonies (in boxes), 1903–1946

Pardess	Season	No. of boxes
	1903–4	22,525
	1904–5	33,479
	1905–6	32,714
	1906–7	62,788
	1907–8	70,546
	1908–9	62,014
	1909–10	71,487
	1910–11	95,262
	1911–12	124,437
	1912–13	175,319
	1913–14	224,371
	1914–15	104,934
	1915–19	—
	1920–21	183,320
	1921–22	237,217
	1922–23	191,358
	1923–24	251,870
	1924–25	251,000
	1925–26	311,000
	1926–27	233,000
	1927–28	340,000
	1928–29	340,000
	1929–30	320,000
	1931–33	No data
	1934–35	1,700,000
	1935–36	1,600,000
	1936–37	2,976,000 *
Pardess Syndicate	1938–39	3,500,000
	1939–40	7,588,181
	1940–41	—
	1941–42	600,000
	1942–43	1,300,000
	1944	2,750,000
	1945	4,500,000
	1946	6,000,000

* By comparison, in 1936–37, Tnuva-Export (the export organization of the Federation of Labor) exported 538,000 boxes.

Compiled from A. Feldman, *Fifty Years of Pardess,* in Hebrew, Pardess Jubilee Publication.

After many hesitations, the unification was effected in 1938, by establishing Pardess-Syndicate. This union, which functioned in the spheres of marketing, supply of materials, and supply of credit, included some 85 per cent of Jewish grove-owners; Tnuva-Export (the cooperative of the workers' sector) and a few new cooperatives remained outside. In order to strengthen the private sector's independence, a supply company (1942) and a transport company (1945) were established.

All in all, then, the story of the farmers' association is essen-

Table 21. Value and share of citrus in Jewish agricultural products, 1927 and 1937 (in thousands of £P on the basis of 1927 prices, and in per cent)

Products	1927		1937	
	£P	Per-cent	£P	Per-cent
Fruit (except citrus)	180	14.0	129	3.4
Field crops	270	21.1	472	12.5
Dairy and poultry	230	18.0	695	18.3
Sundry	110	8.6	228	6.0
TOTAL	790	61.7	1524	40.2
Citrus	490	38.3	2267	59.8
GRAND TOTAL	1280	100.0	3791	100.0

Source: D. Gurewitch and A. Gertz, *Jewish Agricultural Settlement in Palestine*, in Hebrew, (Jerusalem: General Survey and Statistical Abstracts, Department of Statistics of the Jewish Agency for Palestine, 1938).

tially one of economic activity and success, with the pride of place held by the development of the citrus branch. How impressive this development was, and how much it contributed to general economic growth, may be illustrated by a few statistics (see Tables 20 and 21).

The organization of the Jewish farmers in Palestine grew out of a population that lacked ideological drive and missionary zeal, was initially hampered by a government that discouraged political activity and autonomy, and for a long time depended upon philantropic aid. In consequence, this organization was characterized by a specific nature and a narrow scope: its orientation mainly aimed at improving security, credit, cultivation, marketing, and other services; it was internally weak and had no significant linkages with parties and other centers of power. In so far as political aims or channels did exist here—mainly in the Judaean association—they were by and large defensive and responses to the initiative and pressures of other groups.

The farmers were able, even though deprived of most of the public funds, to create and develop cooperative tools which not only helped achieve high productivity but pioneered and sustained the most important economic branch of the country—citrus. They were less geared, motivationally or structurally, to take part in and shape processes of institution building; their share in and impact upon agriculture and the agrarian fabric steadily declined. This institutional marginality was, inevitably, even more pronounced in noneconomic spheres. Culturally and socially (except for the labor struggles), the farmers' organizations represented or supported no ideology and no activity, either vis-à-vis the members or in general terms. Politically, they served primarily local or sectoral economic aims and did not participate significantly in, and often deviated from, national targets, policies, and frameworks. These characteristics will become much more pronounced when contrasted with the organization of the labor sector, presented in the following chapters. However, its two main features stand out clearly: it is essentially particularistic and representative of economic interests rather than of social ideas and a way of life.

6

Growth of the Kibbutz Movement: Hakibbutz Hameuchad

This chapter examines the background and the development of the organizational framework of Hakibbutz Hameuchad, focusing on its structure, on the orientations of its founders and leaders, and on its activities and constant growth. Two distinguishing features of this organization are emphasized: first, its powerful expansionist drive, with the desire to establish a broad social and political system; and second, its assumption of a large variety of general tasks, national and collectivistic. In this context, a long survey of facts and developments is necessary, in order to trace the relative place of the organization described within the political and economic centers of the Jewish community, and in particular in respect to institutions and organs of the workers' movement. An analytical integration is provided in the final section.

The discussion relates chiefly to the period 1905–1944, but some remarks on later processes are appended.

First Period—Historical Background (1905–1923)

The period from 1905 to 1923 was, for the most part, characterized by an almost complete absence of formal suprastructures or institutions that could have united the agricultural settlements organized on collective lines. The few kvutzot founded in Palestine during this period were, as a rule, the result of the initiative of the kvutza members themselves, and no political or social body gave them guidance or direction. True, most of the kvutzot members had the common bond of belonging to certain political parties or movements; but this bond existed only through the central political organization, which actually functioned apart from the periphery.

The peripheral settlements, as such, had no central organization of their own. Even when some signs of a wider organizational activity appeared, their declared purposes and methods of operation were vague and indefinite. Only at the end of this period was the first countrywide organized body created—the Labor Battalion.

Ein Harod was also on its own, and its channels of communication with the center were individual. Even then, however, the quality of active and direct participation in this center, and of nurturing and utilizing connections with it—which was to characterize the Kibbutz Hameuchad organization later—was already evident. We document it by first delineating the structure and policies of the relevant central bodies and then by tracing Ein Harod's relationships with them.

Although the sovereignty of Palestine was vested in the British mandate, and all community institutions were, at least formally, under its authority, there was a wide area of internal autonomy, covering, among others, rural settlement.[1] It is with internal Jewish institutions, therefore, that we are mainly concerned in the present context. Those relevant to our theme belong essentially to two frameworks: world administration of the Zionist movement, and its organs in the country; and the workers' movement, its political parties, and its General Federation of Jewish Labor, the Histadruth. Let us take up each of these broad frameworks in turn, recalling to mind the general background described in our introductory historical survey.

After the occupation of Palestine by British troops in 1917, the Zionist Executive was reluctant at first to resume the settlement activities it had engaged in before the outbreak of the war, because the legal status of the country was not yet clear. Besides, it was inclined first to repair the ravages of the war and to rehabilitate the existing settlements. The Zionist Executive also opposed large-scale immigration during the period 1918–1920 because of the unemployment in the country and the lack of money required for the absorption of newcomers. In view of this, it was decided to establish a special fund for immigration and settlement, the Keren Hayessod (Foundation Fund). The Keren Kayemeth, it was decided, would restrict its activities to the acquisition and reclamation of land.

During this period, the Zionist Executive, through the Vaad Hazirim (the Zionist Commission, established in 1918 to represent the Zionist Executive vis-à-vis the British authorities), also devoted a large part of its funds to the establishment of the public services in the Jewish community, particularly in education and health. As a result, only a small proportion of the available funds was channeled to settlement projects. The settlers and those advocating settlement protested this decision.[2] Thus, tremendous pressure was exercised on those delegates to the Zionist Congress who, within the framework of the labor parties, represented collective settlements.[3]

This pressure was probably one of the principal reasons, if not the only one, for the revision of Zionist Executive policy with regard to agricultural settlement. It was decided to give greater economic assistance to existing settlements, but this time, at least formally, on a different basis than was customary. From now on the settlements themselves, and not the settlement institutions, would bear all losses; but in order to make them self-supporting, they would be given all the necessary initial equipment and supplies, a method that had never before been used. The Zionist Executive now laid particular emphasis upon consolidating and developing its agricultural settlements and diverted its major efforts to those objectives.

Nevertheless, the struggle for the allocation of funds and the recognition of the importance and efficiency of the communal forms of settlement was far from over. Clashes increased with members of the Zionist Executive residing abroad, while resident representatives showed much more understanding. The recognition of their needs and demands was thus a major concern of the settlers and the potential settlers in question.

In the absence of central organs of the collective sector during this period (particularly at the beginning), the major burden of dealing with the settlers' claims fell on the network of institutions that was being developed by the small group of workers then in the country.[4] These were the organizational instruments formed by individual labor parties and by mutual cooperation through an over-all body. We cannot, of course, do justice here to this development, and only its major steps in the period after the First World War are recorded.[5]

In 1918 most of the workers in the country were affiliated with two parties. The first, Achdut Haavoda (United Workers), was the larger of the two; the second was Hapoel Hatzair (Young Workers).[6] These two parties were thus mainly responsible for the establishment, in 1920, of the General Federation of Jewish Labor in Palestine (the Histadruth).

In 1920, at its first convention, the scope and character of the Histadruth activities were defined, primarily under the influence of its majority party, Achdut Haavoda:

1. Establishment and development of agricultural settlements and other branches of work in villages and towns; formation of labor groups and brigades for agriculture and industry.
2. Acceptance of work and its execution (contracting).
3. Organization of labor in general trade unions including all workers in accordance with their trade on a non-political basis.
4. Improvement of working conditions and raising the productivity of labor.
5. Education for work and occupational training.
6. Provision of supplies on a cooperative basis.
7. Medical assistance (*Kupat Holim*—Mutual Sick Benefit Fund).
8. Organization of defense and guard services.
9. Reception of immigrants arriving in the country and work placement.
10. Encouragement and organization of the immigration of workers from abroad.

11. Inculcation of the Hebrew language among the workers.
12. Creation of cultural institutions for teaching general subjects, agriculture, and trades.
13. Publication of trade periodicals.[7]

As can be seen, the activities of the Histadruth were, from the very beginning, far different and broader from the traditional programs of trade unions as we know them in Europe, and even more so from those in the United States. In order to achieve all these goals, a wide network of institutions was established early in the 1920's and consolidated and developed during the following decade.

As stated above, the Zionist Executive, which assumed responsibility for settlement, was constituted on a democratic basis. It was elected by the World Zionist Congress, which assembled biennially. The delegates to the Congress represented political parties and movements that reflected all the political ideologies existing at that time among Zionists all over the world. The influence of the labor parties, which were organized on a worldwide basis, was already considerable at that period, and they were sufficiently strong to obtain one or two seats on the executive body. There they played a double role, representing the interests of their members vis-à-vis the Executive, and dealing with economic and social developments that are normally the responsibility of a sovereign government.

The absence of such a national Jewish power center made the second role of primary importance, and the task of representation became secondary. It is in this context that the fierce political struggle that ensued and the almost frantic effort to create organizational frameworks must be understood.

Achdut Haavoda,[8] the main protagonist, constituted the bulk of the radical wing of the labor movement; in sharp contrast to Hapoel Hatzair,[9] which was more moderate ideologically and more conservative in its political methods, it sponsored new conceptions of aims and channels of political struggle.

The innovating character of the Achdut Haavoda party was expressed in three spheres: its ways and means of recruiting manpower and securing political support; its methods of manipulating manpower resources; and the scope of its objectives.

The party adopted, first of all, the principle of penetrating as deeply as possible into the major classes of the population; admittance to its ranks was virtually unqualified. This party, unlike the others, contented itself with a minimal common platform and refrained from making extreme demands upon its adherents, such as complete identification with its comprehensive political program or prolonged political indoctrination. While élite groups (such as Ein Harod itself) were its mainstay, the conquest of power and the enlistment of the broadest possible support were considered basic conditions in the crystallization of such select groups and vital for the achievement of their programs.

This approach influenced to no small extent the manner in which the manpower resources were manipulated. The party leaders recruited power units that could be utilized, either temporarily or for long periods, in order to attain various temporary or permanent objectives. This idea of a reserve of manpower, at least partly mobile, which could be posted to various "fronts" in the political struggle, was not new in the labor movement, but the leaders of Achdut Haavoda made a great effort to translate it into reality.

The recruitment of additional manpower reserves of mobile immigrants matched the wide circle of objectives. Agricultural settlement, industrial enterprises in towns and villages, public works, health services, cultural and educational centers, the formation of military defense groups, all these were only some of the manifold activities which Achdut Haavoda considered an essential part of its program, in order to lay down the foundations for a socialist and democratic society in Palestine.

The broad orientation of Achdut Haavoda and the important place occupied in its scheme of things by rural settlement were projected into the activities of the Histadruth. The General Federation of Jewish Labor included several trade unions in the European sense of the word, such as the Union of Building

Laborers, the Clerks' Union, etc. One of the more important of these (which actually preceded the Histadruth by several years) was the Federation of Agricultural Workers. The highest executive body of this organization was the Agricultural Council, and its most important office was the Agricultural Center, or Hamerkaz Hachaklai. This department was responsible for the training of settlement groups, their housing, the preparation of plans for each settlement, participation in preparing budgets for the Settlement Department of the Zionist Executive, and for all matters concerning transportation facilities, building of roads, etc. This center also promoted the organization of professional associations to meet the problems of mixed farming and the constant striving to improve the standard of living.[10] These associations served, first and foremost, as centers for the exchange of professional information; in addition, they dealt with the supply and marketing of some specific branches of the economy, through connections with special marketing institutions.

Another major enterprise of the Histadruth was the central economic body known as the Chevrat Haovdim (Workers' Company). The purpose of Chevrat Haovdim, as defined by its founders, was to organize and develop the economic activities of workers in all branches of settlement and employment, in towns and villages, on the basis of mutual assistance and responsibility.[11] Its authority—at least in theory—was to be very wide:

> Chevrat Haovdim owns all the financial and cooperative institutions of the Histadruth. It establishes enterprises and funds and retains the founding shares of all its economic enterprises. It is empowered to impose taxes, determine the rates of pay in its institutions and undertakings (insofar as a system of salary payments exists), and to fix the prices of products. It coordinates the activities of the various institutions and supervises their management, approves their plans, controls their execution, and directs all their activities for the benefit of all workers.[12]

It was thus designed to be the holding company or central organization for all the independent undertakings of the Histadruth.[13] Chevrat Haovdim was legally registered in 1924, and some of its subsidiaries were organized immediately afterwards. All in all, every economic undertaking already in existence, and those established at a later period, were brought under the control of this body.

A systematic analysis of Chevrat Haovdim is obviously beyond the scope of this study.[14] Here we shall only note that, in practice, management and control from the center *lagged* very much behind the formal possibilities authorized by the constitution. However, it did undertake the handling of all the major needs and claims of the workers, including settlement, and a series of specialized subsidiaries was founded (or incorporated) for this purpose. Of these, the main bodies intended to deal with rural colonization (though not necessarily exclusively) were the following:

Hamashbir, a cooperative wholesale society. This organ had been founded before the First World War, in 1911, to deal with the kvutzot's inadequate working capital and the inexperience of their members in marketing produce. Its basic function was to serve as intermediary between the producer and the consumer (particularly those in the working classes in the towns), and also to extend credit to both parties.[15] It operated at fixed prices, conducting all business at its own risk and fluctuating between profits and losses. Final profits were returned to the customers in the form of dividends, in proportion to the sum total of their annual purchases.[16]

Bank Hapoalim—Workers' Bank, created by Chevrat Haovdim as its major financial instrument. The bank was intended, among other things, to extend credit to the settlements. In the early 1920's strong credit institutions did not yet exist within the Jewish economy. The first agricultural settlements were established with the aid of funds obtained fortuitously from various sources; besides the Keren Kayemeth, the Palestine Land Development Company and even J.C.A.[17] occasionally participated financially in the founding of agricultural kvutzot. But it soon became apparent that the systematic development of agriculture demanded a strong financial institution, capable of extending credit for productive undertakings and guarantees to cooperatives for private loans. "The worker cannot become his own master or carry out his plans as long as the survival of each kvutza is dependent, not on its potential economic values, but on appeals to an endless number of bodies subject to a variety of influences, and on having to defend its budget proposals before various institutions that look upon the whole undertaking as a stepchild, if not something worse." [18]

The manner in which the shares of Bank Hapoalim, especially the one hundred founding shares,[19] were distributed ensured that the workers would retain ownership of the bank and indirectly safeguarded the influence of the representatives of the collective settlements on its board. The ownership of the founding shares guaranteed the social character of the bank. They could not be owned by individuals or by just any legal entity, but only by labor institutions or cooperatives. Actually, all these shares were (and still are) in the possession of Vaad Hapoel (Executive Committee) of the Histadruth; however, the capital of the bank was not obtained from the labor sector alone; the Zionist Executive placed a large loan at its disposal,[20] and its representatives sat on the board of directors during the first years. The settlements were not always satisfied with the bank's policies. During the early years, the bank assisted the Histadruth's contracting firm [21] much more than it did the agricultural settlements, a policy which was later changed.

The *NIR* Cooperative, on the other hand, was founded for the exclusive purpose of helping the agricultural settlements. This institution, created in 1924 by the Federation of Agricultural Workers, was to serve as an instrument for the extension of mutual credits to the agricultural settlements affiliated with the General Federation of Jewish Labor. However, it also proposed to concentrate in its hands all contracts for the lease of land and for loans already made or to be made with the settlement authorities of the Zionist Executive, the purpose being to ensure the legal jurisdiction of the Federation of Agricultural Workers over all the agricultural settlements. This objective was not realized, though, as the national funds refused to agree to the centralization of all agreements in the hands of one body and insisted that contracts be signed with each settlement separately. (Some of the settlers, particularly those in the moshav, also objected most strenuously to this arrangement; and we refer to this subject again in the following chapter.)

We have dwelt at some length on the institutions of the workers' movement, because in the absence of a strong political and economic center, mandatory or autonomous, the scope and intensity of these voluntary activities filled a critical gap and constituted an instrument of immense power.

The drive of the labor sector, contrasted with the organization of the colonies, surely explains the paradox of Jewish colonization in Palestine: why the colonies and the private sector as a whole—which had been first on the scene, possessed most of the private capital, constituted the majority of the rural population, produced the bulk of the agricultural produce, and suited the economic and social views of many members of the Zionist institutions [22]—had little power and influence in settlement policy. The workers' organizational frameworks were geared to two complementary and mutually reinforcing efforts: agressive struggle against potential competitors, and control over its own members, through their dependence on services, representation, and help. Later on, as we shall see, the collective (and to a lesser extent the cooperative) settlements had the distinction of carrying out major national functions in such areas as security, dispersion of population, and creation of employment for immigrants, functions to which they were ideologically and structurally suited. Now, however, it was the organization that constituted the main strength.

We must now study the place of Ein Harod and later of Hakibbutz Hameuchad in this effort. The leaders of Ein Harod and the movement it inspired were, first of all, almost without exception active members of the majority party, Achdut Haavoda, and they were the most enthusiastic supporters and realizers of the political principles formulated by this party. Achdut Haavoda was thus its link, and usually its supporter, in the various institutions and bodies of the Histadruth that functioned in the labor sector in general and in the rural labor sector in particular, and in those which represented the movement in Zionist organizations.

The outstanding example of the influence exercised by the members of Ein Harod in the Achdut Haavoda party—and through it in the Histadruth—was their victory in the bitter dispute with the Gdud Haavoda (see Chapter 3). Indeed, the support of these institutions was all that enabled the kibbutz to withstand the Battalion's attack.

Another example was Ein Harod's demand that the party and Histadruth leaders support the struggle against the Zionist Executive's intention to reduce the settlement budget of Ein Harod, an expression of the Executive's serious doubts regarding the viability of a settlement of the size to which Ein Harod aspired. The members of Ein Harod complained bitterly that the party and the Histadruth leadership did not protect their interests adequately. This lack of firmness was probably due in large part to the weakness of the Histadruth, which had only recently been founded; but the Histadruth leadership in general shared some of the Zionist Executive's misgivings about the future of the large kibbutz. However, Ein Harod's plans ultimately were secured by pressure from the Agricultural Center of the Histadruth, in which Ein Harod had considerable direct influence. This recognition thus enabled Ein Harod to realize its expansion program. It will be recalled that, while in 1923 Ein Harod and its affiliated plugot numbered only 220 persons altogether, further groups were being established and awaited settlement; and it was the support which Ein Harod could mobilize that enabled its demands for additional land, economic assistance, and public support to be actually realized.

The Second Period—The Beginnings of an Organization (1924–1927)

This brief but distinct period may be designated as the period of transition from individual contacts and sporadic beginnings to a formal institutional structure. Unsuccessful attempts to organize all the collective agricultural settlements had been made several times even before the period under discussion, the most important being at gatherings of delegates representing all the agricultural kvutzot in June, 1923, at Degania and in November, 1923, at Bet Alpha. On the latter occasion, it was resolved to form an association (Hever Hakevutzot Vehakibbutzim) of collective settlements and groups, but the resolution was never carried out because of divergent views on the structure of the organization and its policy. The differences related in particular to the extent of the authority of the association over the individual kibbutzim.[23]

Shortly after this failure, Kibbutz Ein Harod embarked upon a separate plan of organization on a national scale. However, it had first to overcome the strong opposition of a minority of its own members in the settlement. As a result of pressure on the part of the Zionist Organization to limit the size of the settlement and the lukewarm support Ein Harod received from the Histadruth, a small group of members advocated the abandonment of the expansionist ambitions of creating a national organization, and the concentration of all efforts on Ein Harod itself. The justification of the expansionist policy was most convincingly expounded by one of the leaders of Kibbutz Ein Harod:

> In striving to become a mass movement, we must realize that employment opportunities in the country are limited. We must daily create new employment opportunities for the masses who will come in the future. We must always look upon ourselves as the vanguard which opens up the way for those who will follow. That is why we are called upon to carry out many difficult tasks, such as penetrating into the moshavot under adverse conditions and securing places of work in the towns. Indeed, it would be easier by far to be a Kibbutz pure and simple, dealing with our own problems only; but we want to reach the difficult places, even those where we may fail. If we do not travel this road, we shall not become a mass labor movement.[24]

The orientation toward a countrywide framework won out, and preparations for establishing the organization were made. At the beginning of 1924, Kibbutz Ein Harod had seven plugot affiliated with it, and the major efforts were directed toward expanding this network. Continuous demands were made on the national institutions to recognize Kibbutz Ein Harod as the guardian body carrying all legal responsibility for contracts signed with the various plugot; the recognition of this principle in many cases saved the plugot from severe economic and social crisis.

To ensure the systematic expansion of the plugot, Ein Harod sent its first emissaries to Europe in order to establish organizational and administrative contacts with the pioneer youth movements recently organized, primarily with Hechalutz movement. This was a youth movement established in Russia at the end of the First World War that in many basic respects shared the ideology and background of the founders of Ein Harod. The importance of Hechalutz in providing manpower is reflected in the following figures: A census of Jewish workers in Palestine, taken in September, 1926, recorded that of 31,836 laborers registered, 8,692, about 27 per cent, were former members of Hechalutz.[25] Although not all of them chose agricultural employment, and some abandoned it after a while, nearly 40 per cent were at the time employed as agricultural laborers, as illustrated in Table 22.

Table 22. Distribution of Hechalutz members in towns and villages as compared with other members of the Histadruth, 1926 (in per cent)

Workers	Villages			Towns	Total
	Private settlements (colonies)	Labor settlements	Total in settlements		
Members of Hechalutz	26.1	13.2	39.3	60.7	100.0
Other workers	16.7	9.7	26.4	73.6	100.0

Source: M. Bassok, (ed.), *The Pioneers' Book*, in Hebrew (Jerusalem: The Jewish Agency for Palestine, 1940).

Furthermore, the number of Hechalutz members in villages was 40.6 per cent of the total number of laborers in the private and the workers' settlements. Classification by sex shows that women tended more to go to villages, because there were no suitable forms of employment in towns and because there were very few women in the workers' settlements prior to the Hechalutz Aliya. This greatly facilitated the admission of new women into these settlements during this period.[26]

Finally, it should be noted that, as part of the campaign to recruit manpower, youth movements ideologically oriented toward collective settlement on the land were organized during this period in Palestine. At a later date, these movements proved to be a most important source of manpower for the kibbutzim.

The enlargement of the countrywide network of plugot was by no means easy and not only because of external pressures. No less difficult was the search for both concrete and symbolic expression of the unity of the plugot. Internal discussions on this subject show that an effort was made to apply the principle of the plugot's mutual responsibility in several ways. The first and most important was the organization of economic cooperation. In contrast to a common joint treasury, which had operated during the Gdud Haavoda period, a budget was designed to equalize the budgets of all the labor groups and thus bring about a more or less similar standard of living. This met with tremendous difficulties because the economic status of the various plugot differed, and some of the rich plugot were obliged to support the poor ones. Other means were then tried, especially the centralization of credit.

The equitable distribution of manpower was more successful, and undue competition among the plugot was avoided, because the acceptance of members to the various nuclei began to be guided by the central body. The leaders of Kibbutz Ein Harod and its plugot also tried to concentrate in their hands the allocation of work to the different groups, particularly to those situated near to the colonies and towns. It is worthy of note that in this sphere they frequently clashed with the leadership of the Histadruth and its economic enterprises, because the kibbutz demanded that priority be given to members of the plugot over individual laborers. In this struggle, the members of the kibbutz frequently lost out, as we shall see later on.

The Histadruth leadership was subject to great pressure on the part of the new immigrants pouring into the towns [27] and demanding social and economic integration. The kibbutz leaders complained bitterly about the attitude of the Histadruth institutions, not because they objected to the absorption of these immigrants, but because of their opposition to the policy of the bodies dealing with this matter; in their opinion, these bodies often ignored the proposed system of integration within the collective framework of the plugot. This conflict was part of a larger struggle within the Histadruth leadership: the clash of interests between the claims of the agricultural laborers—or candidates for agricultural settlement—and the urban workers, and the preference of some of the first for cooperative rather than collective principles.

The principle of equality between Ein Harod and the plugot was also expressed in the education tax. The plugot differed in demographic composition; most of them were composed of bachelors and a few married couples without children. But in Ein Harod and in a few of the plugot, there were a large number of children, and their care and education were a great economic burden. The education tax was designed to place the burden of responsibility for the care and education of the children on all the plugot, even those in the towns where this problem had not arisen at all.

To sum up, this was a period immediately preceding the creation of an over-all central organization in the form of Hakibbutz Hameuchad (United Kibbutz), in which the main principles of the movement were forged:

First, Kibbutz Ein Harod received senior status in relation to the other plugot.

Second, the existing plugot, as well as the new ones started by Kibbutz Ein Harod, ceased to exist as small isolated groups subject to constant upheavals because of the high turnover of members and the serious economic crises. The incorporation of the plugot into the framework of the kibbutz gave the members of the plugot a sense of belonging and strengthened their adherence to the central aims that guided the activities of each pluga. Owing to the central status of the Ein Harod settlement, the ideology of its élite became the ideology of the élite of the entire organization. All this helped to develop an orientation toward agricultural settlement among the plugot scattered throughout the country. They all considered themselves affiliated with the body that had successfully met the challenge of establishing a large settlement and was striving to build up similar large settlements in the undeveloped areas of the country. The collectivist orientation of the bodies joining the kibbutz was reinforced. True, they had conducted their affairs on a collectivistic basis before affiliation, but this was the result of circumstances and the difficult conditions attendant upon their integration. The majority had not yet reconciled themselves to the prospect that this form of living should become permanent. Their joining a body that had inscribed these principles on its banner and transformed them into reality in the Ein Harod settlement altered their collectivist orientation from a temporary necessity to a way of life.

Third, the close association between the plugot and the Ein Harod settlement enabled a more efficient allocation of work and of manpower. This enabled Ein Harod to devote itself, more than any other body, to carrying out the major social tasks delineated by the Histadruth institutions. It also inculcated a sense of responsibility toward all labor movement undertakings in the country, and resulted in a close association with the Histadruth and intensive participation in the activities of its various institutions.

During this period, the emergent organization continued to strengthen its participation in the central institutions and intensified its attempts to influence policies and to mobilize resources. As mentioned above, the fourth wave of immigration came to the country during the period under discussion and brought 82,000 new immigrants. The main efforts of the national and labor institutions were directed toward the absorption of these immigrants into the urban or semiurban sector. Most immigrants of that group preferred this sector to agriculture, and therefore the settlement authorities of the Zionist Executive restricted their activities during that period to the consolidation of the existing settlements and diverted their main efforts to the acquisition of urban property and land suitable for citrus-growing. Also, the members of the Zionist Executive abroad, who fixed policy, opposed even more than before experiments in collective settlement.

The Achdut Haavoda party and the political and economic bodies of the Histadruth that grew and developed during this period again served as the main channels through which collective claims were presented to the Zionist authorities. Furthermore, the settlers directly exploited this vast expansion by injecting themselves actively into all the policymaking institutions and by influencing the policy of the settlement department.

In 1927 three members of Kibbutz Ein Harod were also members of the executive committee of Achdut Haavoda party. On the Agricultural Council, the body which controlled the policy of the Agricultural Center, three members of Ein Harod (out of a total membership of eighteen) served, although none sat then in the Center itself. On the board of directors of Hamashbir, which numbered four members, one was a member of the Ein Harod settlement. And while, during the period from 1924 to 1927, no members of Ein Harod or its plugot were yet on the executive committee of the Histadruth, they had six representatives out of thirty-three on the Histadruth council, which met two or three times a year to decide on the economic and political policies to be followed. In order to assess their relative influence, it must be remembered that the membership of Ein Harod and all its affiliated plugot numbered no more, in 1927, than 4.4 per cent of the total membership of the General Federation of Jewish Labor,[28] and they therefore had a higher representation on the council than the proportion of their numbers. The absence of representatives of the kibbutz in the Hamerkaz Hachaklai and the Vaad Hapoel may be explained by the fact that membership on these bodies required a person to devote his full time to administrative work in the town and to abandon agricultural work; at that time, kibbutz members rarely devoted their full time to public service. It may also have been that during this period they had not yet sufficiently consolidated their own organization.

The members of Ein Harod and its plugot affiliates received great encouragement from the Achdut Haavoda party resolutions adopted at its fourth convention in 1924:

> The Convention draws the attention of all laborers to the danger which threatens collective settlements in the form of renewed attacks on it by the opponents of labor settlement in the Zionist Organization. The Convention appeals to Achdut Haavoda party, the General Federation of Jewish Labor in the country, and the Zionist socialist movement in the Diaspora to defend the collective kibbutzim settlements against the many attempts to suppress them, and to increase their share in the Jewish settlement of the country. The Convention declares that the basis and the major inspiration of our economic undertakings in the country, whether in agriculture, industry, or public works, are the organized kibbutzim united by ideology and practice in preparation for a collective form of life in a workers' society.[29]

This declaration clearly states the priority accorded to the large collective settlements in the political platform of the largest labor party of the Jewish population in the country. Thus, Kibbutz Ein Harod attained recognition of its central position and of the type of settlement that was to be given preference by the Histadruth's economic undertakings and services. It was a great and decisive victory by the protagonists of collective settlements over those within the Achdut Haavoda party favoring the moshav form. The bonds between Kibbutz Ein Harod and the Achdut Haavoda party thus grew stronger; but members of the kibbutz maintained the principle of formal separation between the kibbutz and the party; and the view of the leaders of Ein Harod at the time was that relations should be close but not integrated.[30] Although Achdut Haavoda's sanction of the large kibbutz and the organization of a strong countrywide association of kibbutzim actually improved the standing of the representatives and supporters of Kibbutz Ein Harod in all the organs of the Histadruth, in the absence of an absolute majority, they were compelled to struggle for the achievement of their aims with the opposition parties within the Histadruth, and especially with the Hapoel Hatzair party, which was more inclined toward the small and autonomous kvutza type and considered the tendency towards centralized bureaucracy very dangerous.

During this period immigration reached a peak. The immigrants, as noted above, were of the lower-middle class or artisans with no allegiance to the Histadruth sector. It was a period of accelerated growth and organization without precedent in the history of the Jewish community in the country, when the two modern towns, Tel Aviv and Haifa, were built up. The Histadruth, whose major real and potential resources were after all drawn from the urban sector, was compelled to set up organizational apparatus to absorb immigrants primarily into the urban sector. This naturally gave rise to many conflicts around the distribution of available work between the individual workers and members of the kibbutzim; the latter disapproved of the attitudes prevalent in the towns and of the compromises that the Histadruth was compelled to make in dealing with the problem of integration. At the meeting of the Kibbutz Ein Harod council in May, 1925, one of the members thus severely criticized the policies of the Histadruth:

> Not being ideologically prepared or organized, it [the Fourth Aliya] led to a diminution of the stature of the kibbutz movement in the ideological framework and activities of the Histadruth. The political parties have abandoned the kibbutz as the major instrument of labor settlement. Because of the pressure of immediate daily problems, the orientation was shifted to the individual worker, on the one hand, and to the moshav, on the other. Without openly revising its platform, the official circles in the Histadruth began, little by little, to accept the ingenuous contention that it might be possible to build

up our class-oriented organization on the basis of two fundamental ideas: the individual laborers in the towns and the small farms in the villages. . . . This means that cooperative labor groups in the towns are going to be disbanded and that artisans will leave the ranks of the working class.[31]

The leaders of Achdut Haavoda party were not so pessimistic as the representative of Ein Harod and regarded with satisfaction the expansion and stabilization of the Histadruth's economic activities as a whole. Nor was the evaluation quoted above, as we shall see, realistic. The members of Ein Harod, however, saw in the process the slow but sure strangulation of their undertaking. This the dynamic members of the kibbutz could not accept, and they increased their efforts towards internal consolidation, in order to combat the attitudes toward which, in their opinion, the Histadruth leaders had been forced.

It was even more important to combat the hostile atmosphere that prevailed in Zionist leadership circles abroad, owing, among other things, to the increased activity of the right-wing political parties in it. In the final analysis, the labor parties and the Histadruth could not let Hakibbutz Hameuchad struggle against the political parties of the right by themselves; these parties intended not only to give priority to private settlement, but to eliminate the General Federation of Jewish Labor from many spheres of activity that it had undertaken. The labor movement delegates to the Zionist congresses, with the support of some other groups,[32] vigorously opposed the proposals of the right-wing parties to reduce the settlement budget and to renew the Zionist movement trusteeship over agricultural settlement, which would in practice have meant the total disappearance of the kvutza form of life.[33] This coalition blocked the full enforcement of these extreme proposals and, for a short period, even succeeded in getting opposing recommendations adopted. Without assessing the merit of the argument, the crucial point here is surely that the kibbutz succeeded, by stressing the mutual interdependence and common interests of the workers' movement as a whole, to enlist the support of the Histadruth for policies that the latter, to some extent, opposed.

But even the amended recommendations caused great damage to many plugot, and some of them were compelled to disband because of lack of means to continue their activities. The result was that, despite the great burden which fell upon the shoulders of the Histadruth of settling the waves of immigration pouring into the towns during these years, it was compelled to basically reorganize, streamline, and strengthen all its institutions and in particular bodies dealing with agricultural settlement, in order to support the aims it had of necessity endorsed.

In 1926 a change occurred in the structure of Hamashbir. The functions of marketing agricultural produce were separated from the business of organizing purchases and supplies for the agricultural settlements, and were assigned to a new institution called *Tnuva*, a countrywide cooperative for the marketing of the agricultural produce of all collective and cooperative settlements, or, to give a broader definition, of all settlements based on the principle of self-labor. Unlike Hamashbir, Tnuva operated on a nonprofit basis, and the money received from sales was transmitted to each and every settlement. The expenses were covered by the payment of a commission fixed from time to time on a percentage of the total value of the products marketed. Tnuva's capital was raised by shares which the settlements were obliged to acquire; each settlement had to buy shares equal to the gross cash value of the products marketed. The capital was invested in the building of storehouses, dairies, collecting and distributing depots, etc.

However, the creation of Tnuva again tended to impose tight central control over the rural sector of the Histadruth (which had not been very successful in the case of Nir). Thus, the type and quantity of products were to be decided upon by the management of Tnuva; each settlement would send its produce in accordance with Tnuva's instructions; and none would be permitted to market any of their produce except through the new organization.

These controls, as we shall later see, were rationally justified in cases of young and inexperienced villages with a small volume of production. They reflected, at the same time, a basic concomitant of the process of organizational growth: with the growing scope of activity on any level, and the increased dependence of member units upon the center, the latter's dominance over them would be strengthened. Here this process refers to the Histadruth and its component sectors, but later on we shall observe the same process in respect to the movements and their settlements.

After restricting the functions of Hamashbir, an attempt was made to make the business profitable and competitive with private business. But by directing its activities to consumers' needs, Hamashbir lost its social significance. Many found it difficult to differentiate between private business and Hamashbir, as the latter began to cater to potential consumers in all sectors of the population and did not limit its trading activities to the cooperative settlements or the labor sector in the towns. No significant changes occurred, however, in the policies of Bank Hapoalim, except for an increase in its capital by the sale of shares in the country and abroad, and a progressive rise in its deposits.

The Third Period—Crystallization and Growth (1927–1944)

In this period Ein Harod grew from an isolated kibbutz with its few plugot, totalling one thousand persons, to a countrywide chain of kibbutzim, numbering 16,500 persons at the end of the period (of which 9,200 were adults), in forty-five settlements and plugot. The decision to establish Hakibbutz Hameuchad was made in Petach Tikva in August, 1927, after lengthy negotiations with many other kibbutzim. The new organization did not succeed in including all the kibbutzim and kvutzot then in existence in the country, but its council saw in the organization of Hakibbutz Hameuchad "the decisive step toward the consolidation of the kibbutz movement and the organizational framework essential to integrating the immigrant and youth groups into the existing kibbutz camp." The council saw in the new organization "the main instrument for the formation of a single workers' society merging the veteran laborers with the new immigrants originating from various classes and countries, a society which unites them in their pioneering efforts and joint responsibility for the workers' undertakings in the country." [34]

And indeed, while other collective movements were later to emerge and compete, sometimes bitterly, for recognition and resources, an essential brotherhood was maintained that assured the kibbutz its long-held élite position.

The council also drafted the following basic statutes:

1. Principles of Hakibbutz Hameuchad:
   (a) cooperative settlement in communal villages;
   (b) expansion of the Jewish laborers' spheres of work and undertakings;
   (c) the continuous absorption of immigrants;
   (d) social and economic union of town and village laborers, of independent undertakings, and of hired labor;
   (e) support of the Histadruth and its undertakings.
2. Hakibbutz Hameuchad shall organize new settlements and guide those already in existence.
3. Settlements of Hakibbutz Hameuchad shall engage in agriculture, industry, and trade, unite the self-managed settlement and laborer outside, provide as much as possible of their own needs, and cultivate new products for marketing.
4. Candidates shall be admitted as members by each settlement in accordance with its need for workers and in accordance with the candidate's ability to integrate socially and economically.
5. Each settlement shall enjoy a measure of financial and administrative independence. The settlement's plans must be submitted to the higher institutions of the kibbutz for discussion and approval.
6. As a rule, members shall not be accepted by the individual kibbutzim but by Hakibbutz Hameuchad. (Direct admittance of members by a specific kibbutz shall be made only in cases of special social or family considerations.)
7. Hakibbutz Hameuchad shall organize its members, settlements, and plugot for the following purposes:
   (a) kibbutz members shall be available for existing and projected undertakings: formation of new plugot, assistance to settlements during certain seasons, etc.;
   (b) assistance by professional and administrative personnel,

and centralized guidance for settlements and training projects;

(c) the mutual sharing of administrative and professional knowledge and experience by all sections of Hakibbutz Hameuchad.

8. Equality of living conditions, by means of:
   (a) a general budget, which shall make equal provision for the general and individual expenses of the members;
   (b) taxes and funds of the kibbutz (education tax, unemployment tax, responsibility for losses);
   (c) division of the surplus of income over expenditure in accordance with the decision of the center for purposes of investment in its settlements, of extension of vocational training, and of improvement of the standard of living.
9. The highest authority of Hakibbutz Hameuchad shall be the greater council, composed of representatives of the settlements and plugot (every twenty members shall elect one representative), which shall meet once a year. Between the annual meetings of the greater council, the affairs of Hakibbutz Hameuchad shall be conducted by a permanent committee composed of:
   (a) Members of the secretariat of Hakibbutz Hameuchad elected by the greater council;
   (b) Representatives of settlements and plugot (every sixty members shall elect one representative). Members of the permanent council shall be elected to serve for half a year and shall meet once every six months.[35]

Several kibbutzim and plugot followed those which had already joined Hakibbutz Hameuchad at its founding convention. It is worth noting, in particular, the affiliation of part of Gdud Haavoda, from which Kibbutz Ein Harod had seceded about five years earlier. This union became possible only because of a later cleavage in the Gdud itself, which brought it to the verge of dissolution.

An examination of the principal statutes of Hakibbutz Hameuchad indicates that they were drawn from two ideological sources:

From the value system underlying the large kvutza, realized in the establishment of Kibbutz Ein Harod, Hakibbutz Hameuchad adopted as its own the policy of settlement in large units combining both agriculture and industry. It also endorsed self-labor in its own settlements and work by its members as hired laborers in towns and villages. It furthermore recognized the principle of a wholly collective form of organization for its settlements and a readiness to absorb candidates of heterogeneous origin, without meticulous selection. However, it modified the idea of the kibbutz as the only alternative to the traditional town and village, thus recognizing that agricultural settlement was not the only method of achieving national strivings, although it insisted that its principles should be accepted as the *best* way of doing so.

From the Labor Battalion, Hakibbutz Hameuchad adopted the principle of readiness to perform all national and Histadruth tasks assigned to it, by forming a mobile labor force. It also accepted control by a central authority, at the same time safeguarding the autonomy of the settlements on the periphery. By contrast—at least at the beginning of the period under discussion—it rejected the tendency of the Gdud to convert itself into an independent political party. As we have already intimated above, the observance of this last principle was not strictly enforced, and the way was left open for the transition to a new structural form, which will be described briefly at the end of this survey.

On the adoption of the statutes, the formal framework of the central organization took final shape. It would be erroneous to assume that on entering this new phase all the inner tensions disappeared. One of the focal points of contention was the degree of social and economic autonomy to be enjoyed by the settlements or the extent of their dependence on the center. For instance, the policy of the secretariat of Hakibbutz Hameuchad did not always coincide with the aims of the groups of potential settlers. Many of these groups wanted above all to remain as a collective unit and objected to being dispersed in order to meet the current general needs of the existing kibbutzim. The general policy was, of course, to maintain these potential settlers' groups intact, but this principle was often disregarded when the requirements of settlements and plugot made it necessary to scatter these immigrants, at least temporarily. The secretariat's demand for more rational use of the available manpower was often justified. In this way, many kibbutzim were saved from dissolution, and their expansion, which began at this period, was made possible on a much larger scale, especially by transforming the plugot adjacent to colonies (such as Petach Tikva) into permanent settlements.

The immigrant groups of potential settlers became the main manpower reservoir of Hakibbutz Hameuchad. The representatives whom the kibbutz sent abroad worked hard to ensure the success of this project. As early as the founding conference of Hakibbutz Hameuchad, the following resolution was adopted:

> It is the duty of the kibbutz members serving on the delegation of the General Federation of Jewish Labor [whose task was to help organize Hechalutz movement mentioned above] to devote themselves to the preparation of the young generation in Hechalutz movement: to train them for collective life and for manual labor and to educate them in Hebrew culture and class unity.

The founding conference also resolved to form nuclei of immigrant groups to be prepared in training farms and pioneer youth organizations: "these groups, after coming to the country, will continue their association with the movement abroad, will concern themselves with the immigration to Palestine of all members, and will arrange for their employment and eventual collective settlement." [36]

Hakibbutz Hameuchad made certain that this resolution was duly carried out. Their emissaries were among the most active of those sent by the Histadruth, and, indeed, representatives of other labor parties (*e.g.* Hapoel Hatzair) strongly resented their activities. The major complaint against them was that as representatives of a general labor movement, they were duty bound to place before the pioneer youth of Europe *all the alternatives* open to immigrants in the agricultural sector of the country, and not to present their ideology and the forms of organization of Hakibbutz Hameuchad as the most desirable type of settlement and the only way to meet the challenge facing the Jewish community in Palestine.[37] The people of Hakibbutz Hameuchad were adamant. Furthermore, they demanded that the national institutions, which distributed the immigration certificates given by the mandatory government, give preference to members of the pioneer youth movements and argued that "only those who contribute to the absorption potential of the country should be allowed to come. The Jew who is drawn to Tel Aviv because he was compelled to leave—or because the bitter fate of his family impelled him to leave—blocks the way for another immigrant to come; not so the pioneer." [38]

The debate on the criteria for immigrant selection took place not between the parties of the labor movement, but between the entire labor movement and the representatives of right-wing and center parties on the Jewish Agency Executive. The latter argued that giving preference only to young pioneers "would deprive all other Jews of the right to immigrate to Palestine." This argument had no practical basis, since the mandatory authorities allocated only one third of the certificates to "labor immigrants." [39] But the organized strength of this third, intimately aligned with the ideological and organizational outlook of the labor parties in the country, did in fact have a politically disproportionate weight, since the nonlabor interests were compelled to recruit their manpower from among immigrants who for the most part felt no allegiance to any organization or party in the country [40]—a great handicap when they tried to obtain political support.

We have already indicated above the percentage of Hechalutz members in the years 1924–1927 among the labor elements in towns and villages. Statistics regarding the second period reveal that in 1931, members of Hechalutz comprised 84 per cent of the total labor immigration; in 1932, 65 per cent; in 1933, 44.5 per cent. However, in 1933, 57 per cent of Hechalutz immigrants were absorbed in moshavot and collective settlements.[41] In 1937, a general census of Jewish workingmen showed that out of 104,122 laborers, 44,416 (42.7 per cent) were affiliated with Hechalutz or some other pioneer organization. The percentage of Hechalutz immigrants was apparently subject to violent fluctuations. A superficial survey reveals that during the years of economic and political crisis, Hechalutz had a practical monopoly of labor immigration, as there were no other eager claimants for certificates; but in years of prosperity, their share became much smaller. In the years of crisis, 1930–1931, the share of Hechalutz thus reached 80 per cent, but during the prosperous years of 1933–34, its share fell to approximately 40 per cent, or even less.[42]

Even the incomplete figures show that the Hechalutz and other pioneering organizations in Europe continued to be the main reservoir of manpower for the collective movement. From this it is clear why the people of Hakibbutz Hameuchad, who were dedicated to expansion and to continuous development, placed so much emphasis on ensuring the flow of this stream of manpower into the country. The representatives of the kibbutz repeatedly pressed for more recruits. Because of this attitude, they insisted that the basic preparation in the training farms abroad should not exceed two years before emigration, and claimed that the practical training would be provided by the realities of settlement itself.

In certain respects Hakibbutz Hameuchad found the ground prepared for recruitment abroad, because the basis for the youth organizations had been laid long before the representatives of the kibbutz started their intensive activities abroad. However, in Palestine itself, the initiative for the organization of the Israeli youth had originated mainly in Hakibbutz Hameuchad, and here it anticipated the other settlement organizations by many years. Hakibbutz Hameuchad began its activities in two sectors: among the young people in school and the working youth. As early as the beginning of the thirties, the first two settlements composed of former members of the youth movement in the country were established within the framework of Hakibbutz Hameuchad.[43]

The expansion of Hakibbutz Hameuchad's membership and settlements called for the creation of suitable administrative bodies that could direct the economic, social, and ideological affairs of so complex an organization. We have already noted the establishment of permanent institutions—permanent council and secretariat—immediately after Hakibbutz Hameuchad was organized. In the economic field, the secretariat dealt with the organization of administrative committees in settlements, the specialization in various branches of agriculture, the obtaining of credits from banks, and the provision of financial guarantees for the settlements. The secretariat also established a special internal credit fund, which operated on a small scale up to about 1944 on the basis of internal taxation. In 1944 the capital was increased, and the fund was transformed into a financial institution recognized even outside the kibbutz sector.

Much was also done on a very broad scale in the field of political and ideological education. The liberal principles governing admission of settlers resulted in considerable heterogeneity and made it much more necessary than in any other system to socialize the various groups to the general framework and living conditions, and thus to combat potential separatist tendencies. It is in this context that the decisions made during Hakibbutz Hameuchad convention in October, 1936 on intensive cultural and ideological activities among members will readily be understood. The following selected paragraphs quoted from the resolutions adopted illustrate the aims of the program and the detailed instructions provided for its implementation:

> Each settlement of the movement will create a rich cultural life and supply the various needs of individuals, groups, and society in general; this will be done by improving the knowledge of the Hebrew language and culture; by inculcating the values of the workers' movement, and engaging in ideological and educational activities, and by providing books and newspapers, organizing artistic per-

formances, and encouraging the creative expression of the members. Thus, a culture rooted in Hebrew, the Holy Land, and socialist ideals will emerge, integrated with the life of the community as a whole, and with the labor movement everywhere.

Each kibbutz will elect, from among its members, a cultural committee, which will organize cultural and ideological activities.

Ideological activities will be systematically planned and include all members. They will be carried out in groups organized according to the members' ideological-cultural level and will incorporate intensive reading and individual study.

Once a year, a concentrated month of study, devoted as the case may be to the principles and problems of the kibbutz, to the labor movement, to literature and art, and to agricultural and vocational training will be centrally held.[44]

As may be seen, the resolutions provided for communicative activities both within each settlement and by the movement as a whole. In time, these latter grew into a permanent seminar, designed primarily for the training of cadres. The seminar conducted dozens of regular courses during the year, attended by many hundreds of members.

In many respects, this work might perhaps be considered a duplication, as Achdut Haavoda party and, from 1930 onwards, the Israel Labor party (Mapai) [45] devoted much of their efforts to the same intensive indoctrination. But the leaders of Hakibbutz Hameuchad considered that a special forum for communicating its cultural, ideological, and political views was necessary, because the political party was of a looser and more diversified character than Hakibbutz Hameuchad and therefore the scope of its communication had to be more varied; what was more important, the party's views were not always identical with the ideas embodied in Hakibbutz Hameuchad. There is indeed no doubt that these seminars crystallized the distinctive ideology of Hakibbutz Hameuchad, which later led to the secession of most of its members from the party.

In addition to the permanent seminar and various courses held at regular intervals, two other communicative frameworks, with ideological and cultural contents, were established. At the beginning of 1924, a monthly magazine began to appear, as well as periodicals dealing with the problems of education and the young generation. A book-publishing firm was founded in Ein Harod and in time became one of the best and largest in the country. To sum up, it is possible to discern in these extensive activities a definite striving towards economic, social, cultural, and even political autarchy. One of the leaders of the kibbutz wrote about this in 1938: "Our major concern should be to concentrate and accumulate forces for settlement and potential forces for internal organization . . . this is even more important than to rely on national capital." [46]

The years between 1928 and 1944 were, however, much more significant for the national commitments and position of Hakibbutz Hameuchad, which increased greatly in scope and intensity. Several momentous events occurred during this period both in the political and economic spheres. We cannot describe the full scope of these events in this survey, but can only allude to the major turning points relevant to our study.[47]

1. In the 1930's, the clashes with the Arab national movement reached new heights. As we have already noted above, the security situation greatly influenced the national institutions to give priority to the kibbutz type of settlement. With the intensification of the conflict between the Jewish and Arab inhabitants, the mandatory government tried, in 1937, to impose a political solution on both sides, in the form of a division of the country into two sovereign states. The plan failed primarily because of Arab opposition, but there were also many who opposed it in the Jewish community, especially the leaders of Hakibbutz Hameuchad. After the failure of this plan, a White Paper was published in 1939 which was designed to keep the Jews a permanent minority in Palestine: it imposed severe restrictions on land transfers from Arabs to Jews, fixed a maximum quota of Jewish immigration (75,000 Jews between the years 1940–1945), and, on the filling of this quota, a total cessation of all further immigration.

2. While in the years 1928–1929 there was a brief economic

crisis, the period of 1930–1939—that of the fifth wave of immigration—marked a take-off point in intensive economic development of the Jewish sector as a whole.[48]

3. At the beginning of the 1930's, representatives of the labor movement, primarily those representing the Israel Labor party (Mapai), joined the Jewish Agency Executive and assumed key positions on it, such as chairman, treasurer, and the head of the political department.

4. The Second World War first cut off, and later completely destroyed, the large reservoir of people in eastern and central Europe. This compelled the Jewish community, and particularly the kibbutz movement, to arrange for the maximum mobilization of the manpower and capital of the Jewish population in the country, a task which had been somewhat neglected before.

In the light of these developments, there was a change in the policy of the settlement authorities. Until 1935, the national institutions had concentrated mainly on the economic development and consolidation of the established settlements and on the acquisition of land on the basis of its suitability for settlement and its cost. These activities were financed by Keren Hayessod whose contracts with the settlers stipulated that the capital invested in the collective and cooperative settlements was not a grant but a loan, and that the principal and interest were repayable to the settlement authorities. Keren Hayessod even created a special company, the Palestine Corporation for Agricultural Settlement, whose main task was the investment of funds in economically stable settlements, and only with first-class guarantees. With the deterioration of the security situation, most of the financial and the agrarian considerations that had served as the basis for land-acquisition and investment policy were discarded, and strategic and tactical necessity became the predominant factor. The national institutions inaugurated a drive to establish settlements on each parcel of land owned by them in all parts of the country, even if this meant the founding of settlements in isolated areas without preparing agricultural plans in advance, a challenge which the kibbutz was best suited to meet.

Parallel to this, the presence of labor representatives on the Jewish Agency Executive brought about a change in the relations between the settlement authorities and the settlers. The Mapai party's representatives now became directly responsible for settlement policy, and all demands were submitted directly to them.

Beginning with the 1930's, Hakibbutz Hameuchad representatives operated on a broader scale, and their public activities became more pronounced. This was particularly noticeable in the citrus settlements to which the plugot were assigned and in whose vicinity they settled. The economy of most of the collective settlements was based on the earnings of their members as hired laborers in the colonies; this explains the great interest taken not only in promoting the specific needs of the members of the plugot there, but also in the general conditions of all laborers. In many instances, members of the kibbutzim were the chairmen of the local labor councils, and they also participated in many other institutions. Because of the militant nature of their struggle, the members of the kibbutzim more than once clashed with the leaders of the Histadruth and their representatives on the national institutions.[49]

There is no doubt that in this field, too, the representatives of Hakibbutz Hameuchad were the instigators of the conflict and its objectives. The movement considered the right of employment in colonies one of the major issues in the struggle between the labor sector and the right-wing sector. It also regarded the colonies as large centers for the absorption of immigrants and as reservoirs of manpower for permanent settlement within the collective framework. It therefore insisted on the establishment of kibbutzim adjacent to them [50] as vantage points to monitor their development.

Its public activities were not confined to the periphery alone. The representatives of Hakibbutz Hameuchad continued their efforts to ensure that the character of the Mapai party, the Histadruth, and the national institutions should be influenced by the movement's policies. This aim was not always easy to achieve, but judging from the final results, it was quite successful, as is evident from the number of representatives of the kibbutz on the various central organizations.

There was no representative of Hakibbutz Hameuchad among the members of Mapai on the Jewish Agency Executive. But in Asefat Hanivcharim (Elected Assembly of the Jewish Community) Hakibbutz Hameuchad was well represented at the end of this period within Mapai.[51]

Only eight years after the formation of Mapai in 1930, two members of Hakibbutz Hameuchad were elected to the seven-man party secretariat. In 1941, the secretariat of Mapai was enlarged to twelve members, four of whom were from Hakibbutz Hameuchad. The significance of this achievement can be better judged if it is recalled that we are speaking of a united party that grew to great proportions in comparison with the rate of growth of the movement itself; it testifies to the weighty influence of Hakibbutz Hameuchad and the high calibre of its élite.[52]

At the end of the 1930's and the beginning of the 1940's, representatives of Hakibbutz Hameuchad made their first appearance on the executive committee (Vaad Hapoel) of the Histadruth: in 1939, three out of thirty members, and in 1942, ten out of fifty-two; and this was so while the percentage of Hakibbutz Hameuchad members was then (1940) only 9.8 per cent of the total membership of the Histadruth.

Most important, however, was the prominent part Hakibbutz Hameuchad began to play in security affairs in 1930. True, many members of Kibbutz Ein Harod and its affiliated plugot had been active in this sphere before. But the peak of their activities, which had far-reaching political repercussions, was the holding of key positions on the staff headquarters of the Hagana,[53] and later on the staff headquarters of Palmach.[54] In 1939, out of three representatives from the labor sector of the headquarters of the Hagana (constituted on a parity basis: three from the labor section and three from citizens' sector), one representative was a member of Hakibbutz Hameuchad who, in the course of time, occupied the senior post on the defense command. This proportion of one to three was retained until 1948, the date of the establishment of the State of Israel and the formation of the Israel Defense Force.

On the other hand, the Palmach command, rank and file, and home bases were largely in the hands of the kibbutzim affiliated with Hakibbutz Hameuchad. In 1942, out of twenty-eight settlements which served as bases for these units, seventeen belonged to Hakibbutz Hameuchad; and in that year, 41 per cent of all recruits to the Palmach also came from this movement (out of the total of 842 then serving).[55] At a later stage, this numerical dominance diminished, because the Palmach expanded rapidly, but the proportion of members of Hakibbutz Hameuchad apparently never dropped below 30 per cent.

As can be seen, Hakibbutz Hameuchad dominated, at least temporarily, in security affairs, to which its leaders attached political and social importance of the first order.[56] This primacy extended to policymaking and command positions; to the manning of most settlement outposts (by virtue of the establishment of kibbutzim in strategic places); and to the mobilization and maintenance of the élite corps.

During this period, Hakibbutz Hameuchad also continued to concentrate in its hands a significant number of key positions in the central economic institutions of the Histadruth. We find that in 1931 there were three representatives of the movement out of a total of eighteen members in the Agricultural Center, on which it had not been represented before; in 1939, there were already six members there out of a total of twenty-five; and this proportion continued until 1948. From the beginning of 1930 until 1948, the proportion of Hakibbutz Hameuchad members in the central administration of Tnuva was about three to fifteen, while in the secretariat of Tnuva it was one to four.

The organization was well represented also on the Hamashbir management: two out of seven in 1931–1932, one out of four or two out of five in the years 1934–1938, and again two out of seven or eight during the years 1939–1944. It should be noted that these were also years of growth and consolidation in the economic institutions of the Histadruth; consequently, the high representation of Hakibbutz Hameuchad in these institutions served well the economic activities of its settlements in respect to rationalization and reorganization of marketing and purchasing, and the extension of working capital and credits.

During this period, Hamashbir indeed improved and increased its services and was better able to supply a large variety of products, later becoming one of the most important wholesalers in the country. The agricultural settlements, though, remained its principal customers: in 1944, out of a yearly turnover of £P 2,800,000 in sales, the agricultural settlements and their undertakings contributed £P 2,000,000.[57] The vastly increased scope of economic activity and the pressure of settlement representatives also accounted for intensified activities on the part of Tnuva as the central marketing agent of agricultural produce.

Tnuva's first problem was, it will be recalled, the need to choose and plan the various categories of produce to be raised. This task was now essential because of frequent surpluses and the consequent fall in prices; its relative success relieved both the mandatory government and the national institutions of the Jewish community of the responsibility.

The second problem arose because Tnuva was obliged, under pressure from the growing and expanding settlements,[58] to develop new internal and external markets for the produce raised by the cooperative settlements, in competition with the marketing firms belonging to the private Jewish sector and the Arab sector. The costs of production in the private Jewish sector (and certainly in the Arab one) were much lower than in the settlements associated with Tnuva, because in the private sector cultivation was largely based on the employment of cheap nonunion labor. Furthermore, the private colonies grew and marketed special categories of crops in which they had specialized for scores of years (for example, citrus fruit), while the collective and cooperative settlements raised and sold chiefly a large variety of vegetables, in much smaller quantities. Were it not for the over-all control of Tnuva, competent and authorized to fix, at least partially, the size of the crops and the varieties of vegetables to be raised, and to control the current credits, the young settlements of that period could not have competed successfully with the other two sectors. The stabilization process in the settlements took a long time, but at the end of the period under discussion (1944), the settlements associated with Tnuva were supplying 70 per cent of the produce of the Jewish agricultural sector.[59]

No less dramatic was the expansion of the financial organs of the Histadruth serving its agricultural sector. Bank Hapoalim, which commenced to function with a small basic capital of £P 30,000, had increased it by 1944 to £P 310,000, on which it paid dividends of 4½ per cent. The principal clients of the bank were the cooperative and collective settlements, especially those recently established. In 1937 the settlements had received only 32.6 per cent (£P 154,920) of the loans issued, but in 1944 they had already received 65.4 per cent (£P 66,483) of the total.[60]

But this was not the bank's only contribution to the economic stabilization of the settlements. It was very active in the issue of shares to provide capital for several financial institutions whose specific function was to extend credits to agricultural settlements. In this connection, Nir in particular should be mentioned. This cooperative, it will be recalled, started to function in 1924 and was designed to provide long-term loans, but for many reasons it did not succeed, over a long period of years, in becoming a major factor in this sphere. In 1934 a subsidiary company was formed, and this company, in contrast to the principal one, which was a cooperative society, was registered as a shareholding company. The company succeeded in accumulating a capital of £P 1,000,000, from the sale of shares and debentures, and it started to provide long-term loans to agricultural settlements (for seven years, on the average) at 6.5 per cent interest.[61] Representation on this company was therefore important for all types of settlements affiliated with the Histadruth, and Hakibbutz Hameuchad members sat in the management, although in a smaller proportion than in other institutions.

The period up to 1924 was characterized, on the whole, by the absence of any organization or central body for the coordination and representation of collective settlements. The bond that united the kibbutzim was their direct affiliation with a political party in common with other labor groups in the urban sector. It was only towards the end of this period that a common identity and framework began to crystallize.

During this time—in particular in the early twenties—a diverse network of institutions, financial instruments, and organs devoted to problems of settlement was created. This development thus represents an almost simultaneous and balanced growth of the apparatus for dealing with settlers' problems and claims, on the one hand, and of the organization of these settlers for submitting their claims, on the other. This complementary process did not result, needless to say, in a smooth solution of all difficulties, mainly because the pressure of the kibbutz already began to be exercised over and beyond these claims. But the very existence of this "match" between central development and internal crystallization served as a mutual bond and allowed the kibbutz to get in, as it were, on the ground floor of the institutional building. The basis of the future organization, characterized by its high participation in the political and economic center and successful mobilization of the support and the resources of this center was thus laid here.

The following years (1924–1927) were essentially a transition period in the development of a central organization for the agricultural groups that had linked their future with that of Kibbutz Ein Harod, which served them as a guide to social and economic ideals. With Ein Harod as the center, the internal ideological conflict was resolved, and the shape of the social and institutional structure of the organization soon to arise became clear. The ideology and the organizational plans evolved (among them, the recruiting of manpower in the country and abroad) set the stage for the next development.

The speed at which these internal preparations were completed outgrew the ability and inclination of the central organizations to meet the demands of Kibbutz Ein Harod and its affiliated plugot. A sharp conflict thus arose with the representatives of the national institutions, especially with the Zionist Executive, whose headquarters were in Europe. Furthermore, the general policies, political and economic, of the leaders of the Histadruth itself were frequently unacceptable to the members of Ein Harod, despite the party's recognition of Ein Harod as the guiding model for settlement.

The Histadruth, whose great responsibility involved most of the laborers in the country, was compelled to deal first with the problems and needs of the new immigrants pouring into the country during this period. The leaders of Hakibbutz Hameuchad stringently opposed any tendency to alter the proletarian character of the immigrants and any compromise at the expense of absorbing them into the communal system of settlement. The struggle of Hakibbutz Hameuchad leaders was not crowned by full victory, but they nevertheless achieved considerable success.

It should be remembered that this period marked the beginning of the acute political-economic struggle which developed later between the left-wing sector and the various right-wing groups. In this area, the kibbutz leaders played a special and decisive role, both on the ideological front and in the everyday confrontations, and especially in the struggle for social rights and the right of organized Jewish labor to employment in towns and villages. Militant public leadership and acceptance of responsibilities went hand in hand with increasing internal strength. This process of mutual reinforcement between scope of activity and organizational stature culminated during the formative years of the thirties.

The main period in the history presented (1927–1944) started with the formalization of this drive, the decision to establish the Hakibbutz Hameuchad movement. The constitution adopted that year outlined the multiplicity of the organization's tasks, beginning with the establishment of cooperative enterprises and of new communal villages and ending with the betterment of the

Jewish laborer's economic and social position in towns and colonies. In order to achieve these aims, the leaders of Hakibbutz Hameuchad conducted an extensive and vigorous campaign to guarantee the flow of manpower from the pioneer youth organizations abroad and in the country. This period also saw the creation of internal financial instruments as well as of an educational system to ensure effective channels of communication between the leaders of the kibbutz and the thousands of members scattered throughout the country. These institutions served as a forum to develop an ideology that differed in many important aspects from the ideology of the political party (Mapai) to which most Hakibbutz Hameuchad members belonged. In general, however, the formal position of the kibbutz improved, mainly its relations with the national institutions. Leaders of Mapai were elected to key positions on the Jewish Agency Executive and, in the course of time, concentrated considerable power in their hands in all matters relating to the allocation of funds and the selection abroad of prospective immigrants.

The deterioration of the political and security situation further added to the stature and importance of the kibbutz movement, as this type of settlement was considered to be the most effective way of efficiently utilizing the limited economic resources to develop agriculture and to take possession of exposed and distant stretches of land. Since this movement was also the reservoir and base of the élite military arm of the Yishuv and took active part in shaping the defense policy, it enjoyed, at least for a time, an almost monopolistic position in this sphere.

The members of Hakibbutz Hameuchad, as representatives of the Histadruth (and sometimes in opposition to its moderate views), continued to lead the campaign for the employment of Jewish labor in the citrus plantations of the colonies in the Sharon and Judaea areas. They asserted that this campaign would bring about a change in the character of private capital, which, in their opinion, was antinational and anti-Zionist. By providing organized manpower, the movement also played a decisive role in the creation of industrial enterprises such as the

Dead Sea Potash Works and the Electric Corporation; its political initiative and ideological endeavors nurtured a new generation of leaders, who gradually attained important positions in the ranks of Mapai, the Histadruth, and several of its economic branches. In this manner they succeeded to a great extent in forging the character of the various central bodies and ensuring, more than ever before, the fulfillment of the claims of the periphery.

In evaluating this astonishing success of the kibbutz (and we have dealt here with only one of its movements), several features seem to stand out.

The first of these is certainly the character of the collective pattern itself, as we observed it in the case study of Ein Harod. By virtue of this pattern, secure in the commitment of its members and geared to mobilization of its resources, to overcoming difficulties, and to absorbing change, the kibbutz became most eminently suited, structurally, to the performance of first sectoral, and then national, tasks.[62] No less vital was the nature of its ideology, which reflected transcendental social and national values and incorporated an expansionist missionary drive.

Of special interest, however, seems to be the unique organizational genius of the kibbutz. This genius thrived, to be sure, because the collective was a relative newcomer upon the scene and had to carve out a position in the face of obstacles, disbelief, and even antagonism. Its essential success, however, reflected the combination of the widest possible recruitment, for the express aim of gaining mass support and power, together with the ability to attach and nurture ideological, political, economic, and even military élites. The national organization of Hakibbutz Hameuchad thus expressed the motivation and the qualities to outgrow local, regional, or even sectoral frameworks, and the capacity to exert influence upon the political and economic center, not only in its own sphere, but as regards national goals and policies as well.

However, this important, often unique, position and the consciousness of being the best organized public force on the fragmented social and political scene of the country, also created in the kibbutz as a whole the tendency toward movement autarchy and toward the establishment, as it were, of realms within a realm.

Some Remarks on Hakibbutz Hameuchad after 1944

As we saw in the case study of Ein Harod, and later in the development of the movement, Hakibbutz Hameuchad rejected the tendency to become a political entity and fought firmly the policy of the Labor Battalion in this respect. Indeed, Hakibbutz Haartzi of the Hashomer Hatzair movement—another federation of kibbutzim—was attacked and severely criticized by Hakibbutz Hameuchad spokesmen for having organized itself deliberately as a separate political body. Later, though, signs of political separatism began, as we saw, to appear in this movement too. These trends resulted to a large extent from the crystallization of ideological differences, which we mentioned above, but also for many other reasons.

This is not, of course, the proper place to enter upon a detailed explanation of the combination of factors that caused the split between Hakibbutz Hameuchad and Mapai. We shall mention only that in arguments about the measure of democracy to be enjoyed within the party, and about the tactics and the strategy to be employed in the struggle with the British authorities, the representatives of Hakibbutz Hameuchad (with the exception of a small minority), together with a few urban groups, sided with the opposition to the leadership of the party. The opposition could, therefore, be identified by its criticism of the bureaucratization of the party and by its advocacy of a more militant policy in all matters concerning the dispute with the mandate.

The resolution of the party convention held in 1944 forbade the formation and existence of organized factions within it. This decision, which was aimed first and foremost at the formation of an inner political circle of Hakibbutz Hameuchad and its adherents in urban areas, brought about the split within Mapai and the creation of a new political party called the Achdut Haavoda-Poalei Tzion movement. This event also altered the organizational relations between Hakibbutz Hameuchad and the central political bodies. From an association of kibbutzim that was part of a heterogeneous political alignment, encompassing ordinary workers and intellectuals from all ecological and economic sections of the population, Hakibbutz Hameuchad was transformed into a tight political organization with urban branches.[63] These structural changes were also incorporated into the new constitution, which declared, "A common kibbutz movement ideology is the fundamental requisite for the unification and progress of the collective society. It is the life force and bulwark of defence . . . against the disruptive influences and the interference of hostile political bodies which oppose the fundamental principles of the kibbutz and threaten the unity of its settlements."[64]

The minority rebelled against this constitutional statute and refused to leave the original party. Over a period of six years, the two opposing groups continued to function together. But when the leftist leanings of the kibbutz leaders finally crystallized after the amalgamation of the new young party with Hashomer Hatzair party,[65] the inevitable split in Hakibbutz Hameuchad itself finally occurred in 1952. Some of the settlements, among them Ein Harod itself, were divided between the two contending groups; in some, the people in the minority were transferred from their settlement to another. The union with Hashomer Hatzair party, whose major activities were confined to the agricultural settlements, did not last very long, and Hakibbutz Hameuchad very soon returned to its pattern of 1944, the kibbutz-cum-political party. This pattern began to lose its effectiveness at the end of 1964, when for complicated reasons which we cannot explain in this study, the tendency crystallized toward forming an alignment (not a union) with the original party (Mapai), which had been deserted twenty years earlier. The alignment was implemented in May, 1965.

7

Growth of the Moshav Movement: Tnuat Hamoshavim

The development of the moshav organization, like that of Nahalal itself, has to be followed after 1948 for its story to be understood. The movement changed drastically with independence, and instead of playing second fiddle in pioneering, it now became a vehicle for mass settlement of new immigrants. Our analysis thus unfolds in three parts: 1919–1930, the first steps; 1931–1948, crystallization and growth; and from 1949 onwards —population explosion.

As before, we focus in each phase on two interrelated dimensions, the organizational structure of the movement and its relationship to the central institutions. These institutions themselves, if they have been described before, are in the present context mentioned only briefly.

The First Period—The Umbilical Cord (1919–1930)

The crucial factor about the background and the early history of the moshav movement is its origin within two existing political frameworks and function in reference to them. Nahalal had been created by a nucleus associated with Hapoel Hatzair party; while Kfar Yehezkiel, a neighboring village settled not much later, had been founded by members of Achdut Haavoda. This close attachment of the pioneers of the Second Aliya to established bodies and interests of the workers' movement caused the first villages to identify with and to depend upon these organizations. Moreover, the attachment to two separate hierarchies was compounded by the fact that the parties concerned regarded the moshav in different lights.

While Hapoel Hatzair, which had adopted Nahalal, regarded the moshav type of settlement as one of the supreme manifestations of its ideology, emphasizing individualism and social justice, the more collectivist and egalitarian Achdut Haavoda viewed it as a bastard offshoot, a deviation from its values and a regression to a materialistic and capitalistic way of life. Thus, Hapoel Hatzair fostered its first moshav and took part in all the problems of its growth, whereas the relation of Achdut Haavoda to the new experiment was one of apprehension and suspicion. In fact, the latter regarded the moshav, from its initial phase, as an inferior form of settlement compared with the kibbutz, and the attachment of Nahalal and Kfar Yehezkiel to their respective parties meant that in spite of their common features, problems, and interests, the two were unable to cooperate.

This horizontal division inevitably extended to all newly formed nuclei, which had to align with one of the frameworks. This inability to organize became crucial with the increase in the number of moshavim and the growth of their needs for resources and support.

The proliferation of potential moshav settlers was now considerable. The success of the first two moshavim in overcoming initial difficulties, in developing their farms, and in establishing viable communities influenced many of the Third Aliya workers to organize with the aim of founding villages along similar lines. The formation of these organizations [1] was also encouraged by the central Zionist institutions responsible for colonization, which regarded the moshav as a less revolutionary agricultural village and likelier to survive. This confidence was reflected in the willingness to provide budgets and to ensure a relatively regular and fast capitalization.

Little support, however, came from the central institutions of the Federation of Labor. It will be recalled that during this period special organs were established to deal with the settlement of workers, in particular the Agricultural Center, designed to act as a representative of the various settler bodies within the Histadruth, and as an intermediary between them and the institutions of the Zionist movement. However, though moshavim were then more numerous, the dominant power in the center was that of the kibbutz in line with the generally growing influence of the radical-socialist ideology and the centralist orientation of Achdut Haavoda. In consequence, the new moshavim were unable to gain a sympathetic ear, even those in the party itself.

In addition to ideological differences, scarcity of resources played its part. In fact, the moshavim were beginning to compete for both land and potential manpower, which otherwise would probably have been directed to the kibbutzim. Also, many members of Kfar Yehezkiel and other nuclei had spent some time in kibbutzim and had been unable to adjust to the collective way of life; support of groups partially composed of people who had left the kibbutz might have been interpreted as sanctioning desertion of the kibbutz. Moreover, the establishment of an alternative to the kibbutz was considered liable to present people of a broadly socialist pioneering orientation with a choice between two legitimate types of settlement, one of which did not demand a total change of individualistic values and might consequently be more attractive.

Because of the dominance of the kibbutzim in the Agricultural Center, the demands of the moshavim were not well represented in and by the central institutions of the workers' movement. This attitude was described as follows: "The reception of the members of 'organization A,' in particular by the functionaries of the Histadruth and the Agricultural Center, was curt and unfriendly. The Agricultural Center completely denied the request of 'organization A' to arrange for the training of some of its members, who were called 'householders' in the kibbutzim; and, some years later, at a meeting in Nahalal, Ben-Gurion warned the moshavim and the new organizations that they were standing on the border," [2] that is, their membership in the labor movement was questionable.

In the absence of support from the center, the association between the new organizations and the two existing moshavim, and particularly Nahalal, grew stronger. These settlements were a source of inspiration for them and a practical example of how to build up a settlement. A. Assaf described the activities of the organizations:

> The secretariats of the organizations were bombarded by requests and applications from dozens of individual workers in the colonies, from people who had left kibbutzim and others, expressing their desire to join the organizations. The organizations fully appreciated what a difficult task faced the individual preparing himself for membership in a moshav, both in the agricultural and the social spheres. Not only was every candidate examined carefully, but from time to time a screening was conducted among the members who had been accepted, according to their capacity of adjustment to the society and the degree of seriousness that they displayed in preparing themselves for their future role.[3]

Despite these close connections with the pioneer moshavim, the members of the organizations did not receive their agricultural training in them. The moshavim, on principle, refused to take in workers even for the purpose of training; nor could they bear the financial burdens entailed. Consequently, the members of the organizations worked for the farmers of the colonies. They preferred the field-crop farms of the Galilean colonies, as these resembled the mixed farm that they intended to develop in the new settlements. However, this method of training forced the members to separate, and this isolation and lack of group cohesiveness caused many people to leave the organizations, some of which disintegrated completely. Nevertheless, in the five years that elapsed after the establishment of Nahalal, some ten organizations crystallized, four of which actually established settlements during this period.

The year 1926 proved a milestone in the history of the moshavim, when initial steps were taken to establish a national movement comprising the existing villages and organizations and to create a central administrative organ for this movement.

The main reason for this action, led by Nahalal in conjunction with Kfar Yehezkiel and the nucleus of Kfar Yehoshua,[4] was the necessity to exercise pressure upon the Histadruth institutions. However, several other factors combined to lend it impetus:

The sense of mission and the proselytizing drive. The vision of shaping the Yishuv in its image was much more characteristic of the kibbutz than of the moshav. For while the first embraced the militant ideology of a mass movement and a complete social transformation, the second was much more in the nature of an individual solution to personal problems. However, these problems had been spelled out, it will be recalled, in broadly significant terms, and had embodied basic social issues. The success of the moshav was thus conceived of as a general asset that a sense of duty and of national conscience obliged to propagate and disseminate. This feeling was clearly expressed in the meeting held in Nahalal in 1926 in order to lay down the foundation for a permanent organization of moshavim: "The meeting notes with pleasure that the idea of the moshav is gaining momentum among agricultural workers and recognizes that the great responsibility for the preservation of the moshav idea, its guidance, and its realization in the spheres of work and social life compel us to organize for the purpose of preserving the idea and discussing it among the workers of the country."[5]

The elaboration, consolidation, and maintenance of the moshav pattern. Although the moshav institutions had been clearly provided for in the original blueprint and worked well in Nahalal and Kfar Yehezkiel, there were still many areas, such as cooperative enterprises and municipal services, which were uncharted and which would continue to be shaped and developed. It was also realized that the moshav contained potential problem areas, inherent in its largely voluntary basis, especially in relations between the individual and the group; further adjustments and elaborations might be required in time. The center wanted to establish a permanent constitutional committee, charged with interpreting changing circumstances in terms of the basic values, and with adjusting the pattern accordingly. In addition, the center would provide current regular communication and supervision, so that the scattered settlements would develop in accordance with sanctioned provisions.

Centralistic tendencies. As was mentioned above, centralism was a dominant organizational orientation, especially in Achdut Haavoda party, and it characterized many members of the new organizations, especially among those belonging to the Third Aliya.

In the light of the situation in the workers' movement, it is understandable that the new organization of the moshavim was not welcomed with great enthusiasm by the Histadruth institutions, "which considered it as resulting from jealousy of the national organizations of the kibbutz movements." [6]

However, self-determination being an accepted principle of the Histadruth, it was impossible to deny it to the new framework, and opposition took the form of practical power politics rather than of formal obstacles. At the same time, the foundations laid down at the first conference were not built upon until four years later, in 1930. The reason for this was apparently that the moshavim that originated the idea were then in the initial stages of settlement and fighting a hard battle for survival, while the majority of the organizations were still scattered. A consolidated and sustained organizational effort was impossible, and the whole matter was held in abeyance for a few years, despite the general recognition of its necessity among all members of the moshavim and the nuclei.

During this entire period, however, preparations for settlement continued. New organizations arose, each of them independently organizing its members and arranging for their training. At the end of the period under consideration, which was one of slow growth and considerable turnover and desertion, the movement comprised three hundred members in moshavim proper and three hundred members in nuclei affiliated with Nahalal and Kfar Yehezkiel.

The Second Period—Autonomy and Growth (1930–1948)

This period, starting in 1930 and continuing until the establishment of the State in 1948, is marked by an increase in the moshav population and in the number of the villages and settlement organizations; by the consolidation of the movement center, the establishment of which began, for all practical purposes, in 1930; by differentiation and specialization of the roles and institutions of the new center; and finally, by the broadening of its spheres of activity and influence.

The following figures illustrate the expansion of the movement in the years between 1931–1937: in 1931 the number of its members in the moshavim and in the settlement organizations was 1,000, while two years later it reached 5,000. In that year, 1933, there were 500 farm units in thirteen moshavim, comprising some 2,700 people, the remaining 2,300 members being distributed among twenty-five organizations. Two years later, in 1935, the population reached 10,000, of whom 6,000 were members and the remainder children and youth.

In 1937, thirty-five moshavim and sixteen organizations provided the movement with a membership of 12,000. If we compare this with 1933, we find that the proportion of moshavim to organizations was reversed. While in 1933 there were twice as many organizations as settlements already working the land, in 1937 the number of moshavim was over twice as large as the number of organizations (35 to 16). Thus, we have quantitative evidence of the frantic settlement activity that characterized this brief period.

Quantitative growth seems to reflect the essential character of this second period. The earlier years had witnessed the organization and settlement of a small number of ideologically committed nuclei from among Second and Third Aliya members, who had, moreover, received regular development budgets; now, by contrast, colonization was popular and much less selective, as well as severely underfinanced. It is against this background that the building up of the movement organization must be understood; its main preoccupations were economic problems and the acculturation and control of the new settlers. The reader should not be surprised, therefore, that many of the social problems which were articulated in Nahalal in the forties and fifties, were, in the movement at large, given prominence much earlier.

At the first conference of the moshavim, which took place in 1930, the permanent center of the movement was established. At the beginning it was called the Committee of the Moshavim and comprised a council of twenty-one members, an executive committee of eleven, and an executive secretariat of five members, all located in Nahalal.

The movement was founded on a representative and voluntary basis, which was maintained in the following years. The functionaries were chosen from among the participants of the first conference, all of whom were representatives of the various moshavim and organizations. From the moment they were elected to office, they ceased to serve as representatives of their moshav or organization and assumed the duties of functionaries of the center of the movement. In the beginning, Nahalal still occupied a central position in the movement. However, with the increase in the functions of the movement, its center was transferred to Tel Aviv.

The first conference of the moshav movement proved a milestone in its development, and consequently, it is desirable to deal with it in more detail. In this conference expression was still given, first and foremost, to problems of the moshav pattern as such and to practical questions emerging in settlements that had already reached a certain degree of consolidation. Two basic structural strains appeared: mutual aid and the status of the professionals (nonfarmers) in the moshavim.

With regard to the first, it was agreed that personal voluntarism could not cope with the problem of backward farms, resulting from prolonged illness or the absence of a member of the family. An institutional solution was needed, based on a formalized and an equitable arrangement administered by the social-municipal organization.

The problem of the professionals became more serious with the increase in moshav membership and with agricultural development. The status of the nonagricultural residents within a settlement where agriculture was regarded as a central value and a way of life was, of necessity, uninstitutionalized. While it was impossible to forego their services, they could not be regarded as "realizers of the ideal" in the full sense of the word and constituted a foreign and a marginal group. The ideologically oriented smallholders desired the fullest possible integration of the professionals and were ready to grant them equality of rights and obligations in all spheres of life, with the exception of the agricultural cooperative, which would remain the exclusive sphere of the actual farmers. On the other hand, the professionals, whose ideological orientation was usually not very strong, and whose reference group lay outside the settlement, demanded conditions, pay, and facilities equal to those of their urban colleagues. The conflict between the two contrasting attitudes was apparently fundamental, and no acceptable solution was suggested.

On the technical level, attention was given first to the cultural life in the moshavim. The founding fathers, it will be remembered, had hoped to create a new mode of life, rich in self-fulfilment and expression. In practice, however, this aim was not realized, mainly because of the heavy burden of work, which did not allow the smallholders much leisure for other types of activity. The problem was even more acute with regard to the children, who were considered to need special educational and cultural guidance to shape their human and social image and strengthen their orientation to the values of the moshav and their adjustment to its way of life. A practical solution had to be found, and the introduction of a special educational curriculum and of regular cultural activities was discussed.

Attention was also given to taxation of members for municipal services; rates differed from place to place, and the members wanted a unified system. Last but not least, the urgent need for improving methods of farming and of increasing agricultural output was discussed.

Thus, the new forum at once began to deal with both basic constitutional as well as practical problems. It was still, however, largely a discussion group, clarifying issues and mapping out future work. Nor was it then anticipated that the center would pre-empt, in any other than an advisory capacity, matters under the direct jurisdiction of the self-governing bodies of the moshav or the cooperative organs established by the Histadruth for marketing and supply. However, this limited approach was soon to change; the second conference enlarged the scope of activity and formulated concrete plans.

The beginning of the second decade of moshav settlement witnessed the spontaneous organization of groups of workers in the colonies in many parts of the country, with the intention of settling eventually in moshavim. These new settlers, like their predecessors of the Second and Third Aliya, were experienced agricultural workers; unlike them, however, they had no significant ideological orientation, and their connections with the workers' movement were much looser. Indeed, some of them were not even members of the Histadruth. The new settlers wished to become independent farmers capable of making a living from agriculture, but they had no ambition to create a new way of life or to establish an ideal community. They were also reluctant to colonize unpopulated and underdeveloped areas in order to conquer them; they wished to settle in the secure central part of the country where settlement was not accompanied by special difficulties.

The social differences extended also to specific agricultural background and inclinations: members of the new nuclei did not undergo their agricultural training, as had previously been usual, in the field-crop colonies of the Galilee, but worked and crystallized their plans in citrus-based moshavot in the central area. In the wake of economic difficulties now affecting the country, agriculture in the colonies suffered, and grave unemployment among workers was created. The least vulnerable branch in this crisis proved to be the citrus branch; its success, combined with the pressure of unemployment, gave rise to a demand for the founding of smallholding villages specializing in this branch. This suited the vocational basis of the new groups. Indeed, the idea of a citrus farm, supplemented by a small plot of land near the house for home consumption, became very popular among the workers of the colonies. It was pointed out that a farm of this type did not require a large holding, such as was necessary for a mixed-farm depending mainly on field crops, and that it would thus be viable on a much smaller scale, not exceeding an area of four or five acres.

The second conference of 1933 realized of course that such a unit would differ considerably from the original family farm image. However, in the light of the difficult situation of the agricultural workers, ideological objections were put aside, and it was decided to encourage the new trend and to incorporate it within the moshav movement.

This was the basis for the "settlement of the thousand" scheme, which proposed to settle a thousand workers on citrus farms in the central region. The goal was never reached, for the new settlements faced serious economic, organizational, and farming problems in the first decade of their existence. First, contrary to previous practice, they were not financed wholly by public capital, but depended largely on their own resources, accumulated during hired employment and supplemented by loans obtained from private institutions. These institutions demanded a voice in the selection of the candidates for settlement, and since their single criterion was financial standing, rather than personal qualification and commitment to the moshav way of life, the whole social basis of membership was now changed. In addition, relations with the settler-borrowers were established on the basis of individual contracts; this weakened the authority of the moshav over them and created opportunities for deviating from principles and evading local sanctions.

These settlements were, moreover, founded on the assumption that only the minimum of capital should be initially provided. The plan assumed that the farmer would also secure gainful employment in addition to cultivating his farm, and that he would save and invest progressively, until the stage of full farm development was reached. This assumption proved unfounded, for the majority of the settlers did not develop their farms under the circumstances, and they remained neglected whilst the settlers continued working elsewhere. Furthermore, because of severe unemployment, outside work was not always available, and many of the settlers were thus deprived of both alternatives and unable to earn their livelihood in the settlement.

The most serious crisis was in the citriculture branch, the settlements' agricultural basis. As a result of the outbreak of World War II and the closure of all foreign markets, the villages had to switch over to mixed farming, but the plots of land at the disposal of the settlers were too small for a farm of this type, and consequently they remained dependent on outside work.

These circumstances confronted the movement and the Agricultural Center with a fresh problem: the integration of semiprivate settlements into the ideological and organizational framework of the moshav. The situation of the newly founded villages was expressed by one of the members of the veteran moshavim in 1934:

> The first moshavim were created in an entirely different atmosphere from that existing at present, and in material conditions completely dissimilar to those of today. In the cultural sense, it was as if we then stood at the threshold of the millenium, and settlement was conducted in an atmosphere of national and social vision . . . materially, the first moshavim were founded on national capital, which, it is true, was given sparingly, but since the individual member did not invest his own money but received money via the public channels of the moshav, it was easier to maintain public authority over him. In

the new moshavim, the problem of inequality exists from the very beginning, as a result of different individual situations with regard to savings from work in the past, unequal conditions in present work and in all the financial equipment necessary for the establishment of a farm. . . . Nevertheless, from the point of view of the future of our movement, the fact that a large number of new villages was established, in spite of shortage of national funds, is more important than the fact that inequality exists in these villages.[7]

The Agricultural Center, compelled to find a solution to the problem of these settlements, applied an unusual measure: it formed a partnership with private companies that helped finance the settlement by means of long-term loans. The main burden and responsibility, though, belonged to the movement, and this challenge was accepted. As already mentioned, the people of the Second and Third Aliya were characterized by a desire to spread the message of the moshav to the working segment as a whole. This aim was not weakened during the second period, and the new villages were accepted as a barren field that required thorough plowing. It was also recognized that these settlements were troubled by difficult problems in farm planning, in the choice of suitable crops, and in the administration of local affairs, both cooperative and municipal.

The moshav movement thus assumed the two basic tasks of aiding the new settlements: ideological indoctrination and instruction in agriculture and local government. In this sphere, complete support was received from the Agricultural Center, which was interested in converting the villages lacking in ideological orientation into settlements realizing the values of the workers' movement; unable to achieve this goal by itself, it encouraged the moshav movement to enter the field.

However, it was recognized that in the new situation the moshav could no longer rely on independently and spontaneously created organizations, the sources for which were either drying up or else producing poor settlement material; this evaluation was further influenced by the success of the kibbutz in centralized recruitment, training, and organization of manpower reserves through youth movements in Palestine and abroad. The aim of tapping and organizing new settler groups thus became, at this juncture, an avowed purpose of the movement, together with the economic development and the social and ideological consolidation of existing villages.

The new and weighty problems with which the center was faced enhanced the process of role-differentiation within it. Special institutions, whose main concern was the problems of the new settlements, were set up. Among them were a farm committee and a committee for social affairs. The former surveyed the farm situation of the new moshavim and recommended the best means of developing them; the latter dealt with the specific social problems. The most serious of these, from the point of view of the veteran moshavim who had founded the new center, was the constant deviation from the principles of self-labor and cooperative buying and selling, resulting from the peculiar conditions of the new settlements. The first occurred when a settler worked outside the farm in order to accumulate capital and the farm was cultivated by a hired worker, sometimes himself a member of the moshav; the second deviation was bred by the hard economic conditions, and from the fact that the new settlements were situated in the neighborhood of the veteran colonies, offering direct markets without the mediation of the cooperative organizations of the General Federation of Labor. Furthermore, the deviation from the principle of self-labor, grave in itself, also affected mutual aid, which could not be realized, either on individual or on corporate basis, in settlements where the majority of the holdings were tilled by hired labor. The considerable differences in the economic situation and in the living and working conditions of the members also gave birth to opposing interests, and prevented the formation of social integration and of effective government.

The center tackled these problems by means of discussions and persuasion, since at that time it did not possess any means of enforcing its authority on either individuals or villages. As the economic activity of each moshav was conducted directly with the joint cooperative bodies of the Federation of Labor—Tnuva for marketing and Hamashbir for supplies—the moshavim organization had no direct influence on these processes, nor could it apply economic sanctions. True, the movement now played an important role in the development of all settlements, not only by directly providing help and training; it also provided the only channel of communication with the Agricultural Center. However, this mechanism of control worked primarily in times of crisis, when deviations in the citrus farm were structurally inevitable; it did not provide any institutional safeguards in normal times.

Furthermore, as was noted above, the new center was voluntarily staffed; and although some of the main offices became full-time ones, the majority of the functionaries did their work in the movement on a part-time basis. The great difficulty in finding people willing to take on this additional burden thus prevented the movement from creating efficient organizational tools. While the majority of the new settlements were brought into the organization and its activities and were made to acknowledge, if only theoretically, their basic obligations, deviations in the citrus-belt villages remained numerous.

It was against this background that drastic changes in the voluntary basis of the moshav were formulated and introduced. A series of resolutions proposed at the third conference of the moshav movement in 1935 reflects this tendency. The first of these called for a legal change in the nature of the contract between the Zionist colonizing organs and the settlers. This, instead of being individual, should now become collective, with the holdings registered as the property of the moshav as a whole, thus giving the assembly a legal basis for enforcing its authority. The conference also decided to arrange the registration of all moshavim as cooperative societies "so that all property, both public and private, will be transferred to the cooperative ownership of the moshav committee, to enable the collective to enforce its will, if necessary." [8]

Still another resolution legally compelled every settler to join the cooperative institutions of the village and authorized the council of the moshav to employ any social or economic measures at its disposal against those who refused. Finally, the conference forbade members to work outside the farm without permission from the community and also formally requested them to liquidate any property held outside the moshav.

All the resolutions clearly constituted offenses against the voluntary spirit of the moshav and, as such, were hotly debated and vilified by many delegates. Indeed, none of the decisions calling for *legal* changes and sanctions were implemented or even put forward formally for consideration by relevant authorities,[9] and the practical meaning of the resolutions was thus, in effect, a directive to try and increase the *actual* control of each moshav council over the members, and of the movement over the council. In order that the local bodies should become agents of the movement, through which it could exercise control over the settlers, close relations with each village were maintained, and committee members were pressured through both the stick and the carrot. However, this system proved efficient only where the council actually controlled the settlement, but much less so in the majority of new moshavim, where its authority was weak and the system of indirect influence was unsuccessful.

At the same time, while these measures were only partly successful and halfheartedly carried out, they nevertheless presented a significant steppingstone towards organizational control; and this trend was never again reversed.

However, in the large sense, the problems of deviance and control of the citrus-belt moshavim were a detour from the program of mainline progress, laid down by the first and second conferences, which emphasized constitutional development, consolidation of village economy and institutions, and guided recruitment. And it is to these that we must now turn.

As was mentioned above, personal voluntarism and individual help were recognized early as an insufficient basis for equitable and proper mutual aid to moshav members; and corporate responsibility, on the pattern of Nahalal, was envisaged. In 1935 this principle was formally embodied in a movement decree, providing for welfare taxation and administration in each village. The decision, though, could not be put into practice in most of the new settlements, where the principle of mutual aid as such was weak, and which found it in any case difficult, as was seen above, to establish a viable corporate framework. However, even in the veteran settlements, this kind of mutual aid was insufficient to solve the problems resulting from serious and permanent causes: disability, death of one of the parents, natural disasters, etc. Thus, the early forties witnessed the establishment within the financial instrument of the moshav movement of the Inter-Moshav Fund for Mutual Aid. This fund took care of households requiring financial aid on a scale too large to be provided by local arrangements.

During this period, the farmers' efforts to integrate the professionals fully into the economic and social life of the moshav until they became full members clashed with the tendency of the professionals to safeguard their status and gains without subjecting themselves to the standard of life and the obligations of the farmers. Thus, while the percentage of professionals in the established moshavim reached 35 to 40 per cent, and in the new moshavim 25 per cent, their influence on the social life of the village was insignificant. Indeed, during the whole period their position in the moshav was marginal, and they did not affect, or contribute to, its policies and social development.

The question of municipal taxes was solved in the decade following the first conference, when progressive taxation was introduced in most of the settlements. The new system was characteristic of the movement's methods: it was introduced in a number of the veteran moshavim and transferred to most of the other settlements by a special committee appointed for this purpose, on the basis of discussions and consultations with local representatives.

In the sphere of economic development, considerable progress was made. It was recognized, first of all, that the structure of the family farm and its gradual capitalization schedule left little margin for error and for improvization,—especially since protracted or unsuccessful growth necessarily prolonged and intensified dependence upon outside employment. The moshav, unlike the kibbutz, did not allow any nonagricultural enterprise, and thus had no provisions for supplementing income within its own framework. Tight and careful planning of each stage of development was thus a basic requirement; and an undertaking was secured from the colonizing authorities to ensure a detailed plan and, as far as possible, a regular development budget in each new settlement.

During this period, close cooperation was also established with the central agricultural experimental station; in the areas of planning training, improvement of methods, and crop rotation, the movement became the main channel of communication. In this connection the farm committee grew and specialized, forming many new vocational subcommittees in direct contact and consultation with all the colonizing bodies and encouraging the creation of professional farmers' associations in the main branches of the moshav economy. Of the two functions, planning and current help, the first was compulsory and binding on all the individual villages; transmission of knowledge from expert to settler, on the other hand, was on an optional and voluntary basis. The majority of the instructors came from the veteran moshavim, and their chief contribution was their own experience and not any specific, formal agricultural education. The brunt of these activities was directed to the new settlements, which were in great need of it, since the agricultural training of some of their members was deficient, their farms not yet established, and their conditions different from those of the veteran settlements.

Despite the general recognition, on the part of its leaders and rank and file, of the need for cultural and organizational activity among the youth, the movement contributed relatively little. The following quotation from *Tlamim* illustrates the situation in this sphere:

What have we done for these youngsters? We must be honest with ourselves and answer, almost nothing. The various initiatives that were not continued and the numerous resolutions (the one recommending the establishment of an institute for secondary education) which have existed on paper for many years, but have not been fulfilled, do not add up to action of any consequence . . . up till now we have not created a body capable of uniting these individuals, educating them, and organizing them for settlement. This body cannot be created without effort on our part. Is there not a danger that without our direction many of them will disperse to all parts of the country, drift into different jobs? We, and all those to whom agricultural settlement is dear, should bear in mind what a great blessing this human element is apt to be to the movement.[10]

The youth of the moshavim was organized within the framework of the Histadruth's working youth movement. However, its membership was mainly urban, and it was thus an empty framework for the moshav youth. The conditions of life and the personal aims of the urban working-class youth were so different from those of the rural youth that there was very little in common between them, and they scarcely spoke the same language except on the level of ultimate national values. Unorganized and lacking an active leadership, the moshav youth was unable to influence the youth movement and to communicate its values to it, as the kibbutzim did. In the early forties the moshav youth independently organized. They realized that the general framework did not suit their special needs and that it failed to provide the ideological basis necessary for identification with the value system of the moshav. The moshav youth was also impelled into independent action by the impulse to create something of its own to express its unique character.

However, although a special section for moshav youth was established within the framework of the working youth movement, it failed to become a formative force in the lives of the moshav youth. One of the reasons for this certainly was that its leaders were called upon to fulfil national tasks during the war, in the allied armies and in the ranks of the military organs of the Yishuv—the Hagana and the Palmach—which opened new horizons and new spheres of activity. However, there were also structural reasons in the moshav itself, deriving mainly from the organic integration of the moshav youth into their parents' households and their loose connections with any type of youth culture.[11]

At the same time, the organized moshav youth did create several communication channels, such as seminars, summer camps for different age-groups, and its own monthly journal, *Maanith* (Furrow). They also set up a small number of settlement nuclei composed of moshav youth who could no longer remain on the family farm, but desired to build their future in moshavim. The moshav youth thus adopted agriculture as a way of life, following in their parents' footsteps. However, this continuity did not derive from special education and organized action on the part of the movement or the youth itself, but rather from the early socialization towards rural and agricultural life that the youth received in their families and in the social setting of the moshav. This was possibly the reason why identification with the moshav way of life was particularly strong as regards agriculture and farm development, and much weaker in respect to social values and the self-limitations imposed by them. By the same token, the degree of continuity in the new villages, which were only partly agricultural, and socially and organizationally problematic, was much smaller than in the veteran ones.

Social and cultural activities was another field where much was said and little done. To the extent that cultural activities did take place, they were within the framework of the individual moshav and depended upon the differential abilities of the local people and upon the public's capricious interest. In general, the standard in the moshav was thus lower than in the kibbutz: the way of life of the independent farmer, who was busy working on his farm for long hours, even on the Sabbath, left him very little leisure time to devote to interests outside his daily work. Also, in this sphere, more than in the agricultural one, the movement suffered from lack of volunteers willing to devote their time to public activity. Thus, the declared aim of the movement to raise the cultural standard of its members as individuals and as a body was not fully realized, and its achievements fell short of the ideal of a smallholding élite. The outstanding exception was the periodical *Tlamim*, which appeared regularly after 1934 and fulfilled an important communicative role by providing a forum for economic and social problems of the movement as a whole and of its individual settlements, as well as for the discussion and participation in general issues and ideas.

Another problem was the movement's lack of independent financial means and organs; and although a compulsory membership tax was introduced, many of the settlements—particularly the new and the poor ones—did not pay it regularly. The resulting inability of the movement even to support those in need of urgent credit for basic agricultural equipment worried the leaders, who consequently proposed the establishment of a regular financial instrument to supply loans and help in obtaining outside credit by providing guarantees. In addition to its financial aspect, the fund was also calculated to achieve social and organizational aims, to bridge the large gap between the developed and the underdeveloped settlements, and to provide a means of putting pressure on villages not observing the moshav authority. The fund was established in 1940, on the basis of shares bought by every settlement in the movement, and played a part in providing aid and in helping families whose male members had volunteered for military service. However, the purchase of shares lagged far behind, and the young moshavim joined only slowly and hesitatingly; all in all, the accumulation of capital was slow and the fund only modestly successful. The weakness of a voluntary fund was expressed in an article in *Tlamim* in 1943:

The discussions on the question of the fund were concluded by accepting a minimalistic resolution. But even these requirements seem too demanding to the majority of our members. We have always played down our financial demands from ourselves and therefore always been slow in our actions and achieved little. The moshav member, devoting all his attention to the cultivation of his farm and struggling for his livelihood, regards the economic values he has created as a reward for his labor and as a manifestation of his success. He strives to improve his material situation in two ways: by increasing his income and by lowering his expenditure, and by expenditure he generally means taxes and funds of all kinds, in which money is invested and an appropriate reward is not immediately apparent.[12]

The dilution of the ideologically oriented veteran nucleus by organizations with only a vague sense of commitment was quite early recognized as a potential danger to the movement. In spite of this awareness, however, the movement did not initiate the establishment of organizations or actively recruit for the moshavim.

The reasons for this were many and complex. The fundamental inhibition was the lack of any missionary zeal in the moshav. The veteran members had been highly conscious of the virtues of their idea and had recognized the need to promote it; they were prevented, however, from any active proselytizing and from any expansionist drive involving sharp and aggressive competition. A concerted effort in this direction was also made difficult, on the organizational level, by the fact that the inclusion of the citrus-belt villages, with all their problems, stretched to the utmost the meager resources of the center in manpower, money, and organization; the unanticipated growth thus prevented any planned one. Instead of itself mobilizing and promoting settlement nuclei from among likely candidates, the movement concentrated instead on helping, through its committee for social affairs, any new spontaneous organization in the management of its internal affairs and in the maintenance of social integration and firmness of purpose.

Among these activities, the first was the assistance given in the formation of cooperative work groups composed of the members of the settlement organizations. These groups resembled the work groups of the kibbutzim, but while all the bodies planning to settle in collectives were organized in this form, only a minority of the organizations planning to settle in moshavim accepted it. On the whole it was more common in the distant regions of the country, but not entirely unknown in the center as well, where some of the organizations assumed the collective pattern of living during the period of waiting for settlement. The cooperative group was engaged in various jobs and also maintained a small farm for partial self-supply. Daily life was organized on a collective pattern and the wages of all the members were pooled.

The moshav movement's attitude to this arrangement was positive, since it regarded it as a guarantee of the future observance of the principles of the moshav by those potential members. It also regarded it as a means of achieving social cohesion of the group and an ideological orientation to the movement. The movement sometimes assisted in initiating these groups and always fought for their right to work in the forum of the Agricultural Center, during the period of partial unemployment and keen competition on the part of the kibbutz groups. It also provided institutional legitimation by working out a constitution for work groups, in consultation with the members of the groups themselves.

However, prolonged collective arrangements were beginning to prove a great hardship for people anticipating and preparing for family-based life. The movement thus regarded it as its duty to alleviate this strain, which it considered a danger to the very existence of the organization. This was done by giving them a feeling of belonging to a corporate body that protected their interests and desired to aid them in overcoming difficulties. Representatives of the movement visited them regularly and discussed matters with them; as time went on, these activities developed until active participation in the life of the settlement bodies was institutionalized, before the settlement was established and afterwards. Sometimes the movement sent a regular instructor who remained in the new moshav for a considerable period of time, but at all events, close contact with the new organization and the new settlement was maintained.

No less difficult for the moshav was the recruitment of manpower through the socialization of youth to its pattern of settlement, a system which had been adopted by the collective movement in its several branches. The main obstacle lay in the fact that the moshav required people with basic agricultural experience, capable of at once assuming responsibility over the family farm. In the absence of training facilities, there seemed little sense in bringing over inexperienced youngsters instead of seasoned workers from the colonies. The lack of such facilities resulted, it will be recalled, from the unwillingness to introduce into the moshav, even for temporary training, any hired, non-farm-owning hands (an unwillingness which was not overcome until the coming of German refugees). It placed the movement in an awkward position, because representatives of the collectives worked energetically abroad, with each movement attracting members to its cause by pointing out its advantages over all the other alternatives. Moreover, the kibbutz representation extolled the virtues of collective settlement against other forms, including the moshav. The moshav was inevitably offended by this propaganda system of the kibbutzim; they feared that, because of negative impressions received before actual acquaintance with reality, the members of the youth movements, and in particular of Hechalutz, would not choose the moshav as a suitable way of life. Thus, the movement, for self-defense, entered the competition for manpower unprepared and against its will.

However, difficulties in getting members to assume public duties, in particular duties that entailed leaving the farm, prevented any large-scale effort, which might have burst the existing framework at its seams. As it was, the activities undertaken were in the nature of information rather than recruitment proper, and did not result, as far as we know, in the coming to the country of even a single group defined as a moshav organization.

No less difficult was the entrance of the moshav movement into the local youth scene, which was inhibited not only by the above-mentioned lack of universalist zeal, and organizational difficulties, but also by the practical monopoly of the collectives in this sphere. The hard fact was that the pioneering youth movements of the country were pre-empted by the first-comers and under their exclusive control; some of them (Gordonia and Hashomer Hatzair) were indeed attached to a parent body in the Diaspora that had adopted the idea of the kibbutz and worked for its development. The collectives never had any serious difficulties in recruiting members to act as leaders and educators in the youth movements, and they even initiated, it will be recalled, the movement of secondary-school pupils, in which they had a decisive influence. Furthermore, they achieved control over the general working youth movements of the Histadruth and organized from it groups for settlement in kibbutzim. The moshav was thus faced with a reality difficult to alter, and since it did not possess sufficient resources for effective public action, it made no effort to compete with the collective movements in organizing youth nuclei.

A new and unanticipated source of young membership potential, albeit relatively small, was tapped with the coming of refugees from Germany, including groups of Youth Aliya. As was described in the case study of Nahalal, the decision to absorb them was hard; once taken, however, all veteran moshavim applied for their share. The training and employment was provided on a family basis, with each family taking in one youth, who participated fully in family life and in work on the farm. The absorption project proved successful, both individually and institutionally; the members of the Youth Aliya formed strong ties with the families with whom they lived, and many of them later joined organizations with the intention of settling in a moshav. As the introduction of the young people did not upset the existing social structure and did not require any special organization, this form of absorption continued until its source dried up with the outbreak of the Second World War.

At the beginning of the war, the movement started to take organized action among other new immigrants, though only on a small scale. Even now the initiative to organize did not come from the movement but from the newcomers themselves; but it was a challenge which the movement could not ignore and which it had to support actively. Understandably, dealing with organizations of new immigrants was more complicated than dealing with those composed of veteran members. They had to be taught Hebrew and introduced into modern Jewish culture, as well as trained in agriculture and socialized into corporate life. For this purpose, the movement had to put regular teachers and instructors at their disposal. This it managed successfully, because by this time its organizational mechanisms were better developed than they had been in the early thirties, and because the number of these organizations was small (nine by 1941). However, the fundamental importance of this venture cannot be underestimated, as it laid the foundations for confidently meeting future challenges, in which the mass absorption and the mass settlement of new immigrants became the main focus of the movement.

Immediately after the Second World War, the moshav movement received an increase in manpower from another source of a special character: groups of soldiers released from the British Army. In the course of their military service the soldiers had organized with the intention of building their future in an agricultural settlement, but of a special type: the cooperative settlement (*moshav shitufi*). In this pattern, each family has its own household which it manages independently, as a unit of consumption and of socialization, while the farm is collective property; all work is done by members on an equal basis, and the income is divided equally among them. In the moshav shitufi the degree of egalitarianism and collectivism is obviously greater than in the smallholders' cooperative village, but smaller than in the kibbutz. Unlike the moshav, it recognizes industrial projects as a legitimate branch of production. Superficial speculation, in the absence of empirical data, would suggest that this preference grew out of the peculiar conditions of military life: living and working in common quarters with comrades-in-arms on an equal basis on the one hand, and the desire for a certain amount of privacy on the other. To such potential settlers the moshav shitufi might indeed have appeared the most appropriate pattern of settlement, and the number of new villages of this type established shortly after the end of World War II exceeded the number of moshavim founded during this period, dropping to a dribble afterwards. Any detailed analysis of this type, however, would require a case study of its own.

All in all, then, the second period in the history of the moshav movement was one of constant but moderate growth. This process was temporarily halted during the period of organized Arab attacks from 1936 to 1939. In 1943 there were thus fourteen thousand members and forty-eight settlements in the movement (compared with twelve thousand members and thirty-five moshavim in 1937), an increase of two thousand in a period of six years, part of which was a result of natural growth. However, it would not be correct to assume that during this period only thirteen new settlements were added to the movement. In reality, a larger number of settlements was established in the years preceding the war, but at the same time a number of moshavim left it. These were mainly moshavim established by German immigrants who lacked the socialist orientation to the General Federation of Labor, and whose ideological commitment to many of the basic values of the moshav was weak. These settlements set up their own organization, called the middle-class settlement, and also a separate marketing organization. In addition, a number of settlements desiring a more flexible social and economic framework, free of the moshav restrictions, also left the movement.

As we shall see, these reverses were more than made up later. Not so, however, the fundamental loss of channels for guided and selective recruitment. This was perhaps the most crucial and decisive feature of the whole period discussed. While the moshav has, in time, become numerically stronger and has, with the addition of postwar new-immigrant villages, surged ahead of the collective, it lost here the chance of becoming a leading settlement élite.

A central factor in the formative period of the moshav movement, and cutting across all the specific problems mentioned above, was the relations of this movement to the central social institutions. In the previous chapter we saw the power wielded by workers' representation and organs in general, and the importance attached to settlement, in particular; here the question is the status of the moshav within this framework, in comparison with its most direct competitors, the collectives.

During the second period, the relations between the moshav movement and the settlement authorities were not direct, but conducted by means of the Agricultural Center, the Histadruth institution whose function was to take care of all matters concerning the workers' settlements. In this period the Center grew in stature and became a guiding factor in settlement policy, the more so as the settlement authorities no longer conducted negotiations with individual settlement bodies, but only with this organ acting as their representative. It thus became the intermediary in the distribution of the Zionist movement's funds allocated to settlement; it maintained, moreover, financial resources of its own, to supplement the usually insufficient budget of the Zionist movement. In contrast to the earlier period, when each moshav or organization presented its own claims through two parties, now the moshav representatives in the Agricultural Center, like those of the collective movements, presented their claims centrally. The recognition of the representatives of the moshavim as legitimate representatives of all the settlements in the moshav pattern did not come easily.

It is only recently that the moshavim began to appear as an organized movement, and our institutions have not yet accustomed themselves to this. Everybody is used to regard Hakibbutz Hameuchad, Hashomer Hatzair and Hever Hakvutzoth as united movements, each of which is entitled to speak on behalf of the whole movement. But they are not used to regard the moshav movement as possessing this right, and this is perhaps the main factor which accounts for the stand taken by the institutions of the workers' movement towards the moshav movement. However, despite this, the moshav movement has recently managed to break its narrow boundaries, and at present the institutions regard it as a duty to accept and offer consideration to the representatives of the movement.[13]

This partial recognition was expressed in practice by the low number of moshav representatives on the settlement institutions of the Histadruth: the Agricultural Council and the Agricultural Center. In 1931 there were four moshav members and seventeen kibbutz members (belonging to all the different movements) in the Agricultural Center, whereas in the Agricultural Council the numbers were six and twenty-nine respectively. In 1939 the moshavim had five members in the center out of twenty-five.

There were also essential differences in approach and ideology that prejudiced the attitude of the majority in the Agricultural Center towards the claims of the moshavim. The main point of dispute between the moshavim movement and the majority was the question of settlement priority. In that period many groups organized for settlement, and work groups waited for years in the hope of being given the necessary facilities to settle on the land. The situation required that universalistic principles be set down for assigning settlement priorities to the waiting organizations. The kibbutzim, which recruited most of their members from the youth movements abroad, demanded that the organizational seniority—existence in nucleus form—of the applicant body, in Palestine and abroad, be taken into consideration. The moshavim, whose candidates were recruited from among the individual workers of the colonies, required that the length of stay in the country and the family status of the individual member should be the dominant criterion. After lengthy discussions, the principles put forward by the moshavim were accepted, but in practice exceptions were usually made in favor of groups that defined themselves as collectives. The moshav thus won legal status, but not actual equality.

Another area where equality with the kibbutzim cropped up was employment for groups organized for settlement. These groups, which lived in or near the colonies, worked primarily in agricultural labor in them, and their jobs were often acquired through the Agricultural Center. The question of obtaining work was frequently a matter of life and death for the group, and in this sphere as well there was a tendency to give preference to the kibbutz plugot over the moshav organizations. The moshav representatives in the Agricultural Center defended their rights on this issue too, but not always successfully.

In other spheres, where the question of equality did not arise, essential differences in approach appeared. One of these was that the moshavim demanded that an effort be made to create the best possible conditions for settlement, and that maximum consideration be given to the needs of the individual settler. One of the main problems in this context was that of the areas allocated for settlement. The Keren Kayemeth, which acquired the land for settlement, adopted the policy of buying land in the unsettled regions of the country; they believed in the importance of establishing settlements in all regions. However, there were also security and political grounds for this policy, and finally, there was a financial consideration, too: land in the far-off regions was cheaper. The question of the adequacy of the land for settlement was seldom considered. The kibbutzim did not oppose this policy; on the contrary, their pioneering orientation approved of the conquest of new areas and of developing desolate regions, especially since they could function under such conditions much better than smallholding systems.

On the other hand, the leaders of the moshav movement opposed this policy fervently. The structure of the moshav and the mixed farm made this form much less adaptable than the kibbutz to the development of arid zones and to heroic tasks. The moshav representatives in the Agricultural Center expressed the desire for consolidation of the existing settlements and demanded that land be bought in the center of the country, in the citrus region. In presenting these demands, the moshav representatives were under considerable pressure from the organizations that had established themselves in the citrus region (Judaea and Hasharon) and intended to settle there. The Keren Kayemeth could not and did not want to change its land policy, and since the majority in the Agricultural Center did not support the claims of the moshavim, the latter remained isolated and at times met with a hostility on the part of their colleagues.

The transition period before settlement was another focus of conflicting attitudes between the moshavim and the kibbutzim. Because of the shortage of land and money, this period was often very long, continuing for several years. For previously mentioned reasons, it will be understood that this arrangement affected the moshavim more than the kibbutzim: members of an organization planning to establish a moshav desired an individualistic framework for themselves, and the collectivistic way of life was a burden to them. In many cases the situation resulted in the disruption of the organization, its members having despaired of eventually establishing their homes in a moshav.

To preserve the unity of the moshav organizations, the representatives of the movement demanded that the waiting period be cut down, and they were even ready to pay for this by foregoing the establishment of additional settlements. This attitude contradicted that of the kibbutzim, which was one of "expansion and conquest." In the Agricultural Center the kibbutz policy was implemented, and according to it, the settlement bodies were compelled to wait for a long period before they were allowed to settle. Although this policy made possible the establishment of many settlements under the hardest conditions, there was a price attached: a large number of members abandoned the settlement bodies, particularly in the moshav organizations. This issue thus reflected a basic dilemma in development in general and in agrarian development in particular, that of steady consolidation as against dramatic spurts, in which advances are made as it were without safeguarding the rear; [14] the dominant approach clearly favored the latter. The argument of the moshav was, however, couched in human and individual terms, rather than those of agrarian principle; it reflected the basic insecurity of the incipient family farm:

> In general, the man living on a moshav is a keen worker with initiative who cannot put up with conditions that threaten him with an unbalanced budget or with a deficit. It will not console him even if it is proved to him that he has created a surplus value equal to the funds invested in his farm. The moshav member will not be able to

continue if he knows that after all the efforts of the family his earnings are insufficient to cover his living costs, and that his debts are on the increase, without any clear prospect of building up his farm in a reasonable period of time . . . the kibbutz [in contrast] has found many possibilities of adapting to conditions below the necessary minimum, of maintaining its farm by creating deficits, unbalanced accounts, and by getting involved in debts on a scale that would make the moshav member shudder.[15]

The size of the settlement was yet another issue on which there was disagreement between the majority in the Agricultural Center and the settlement authorities on the one hand, and the moshavim movement on the other. The latter was interested in large settlements capable of providing the necessary services on an adequate level and of maintaining a high level of cultural and social life in the moshav; in order to achieve this, they were even willing to forego the allocation of the necessary funds for rapid consolidation of the settlements and to prolong the period of consolidation. Experience had shown that the capacity of small moshavim to maintain themselves and overcome difficulties was limited, and that these difficulties were too great for the settlers to bear. However, the settlement authorities and the Agricultural Center wholeheartedly adopted the policy of dispersion of the population and the establishment of settlements in all possible locations, even where the local conditions were not suitable for the establishment of large communities. The point of view of the moshavim was rejected.

All in all, the militant attitude of the kibbutzim, who regarded their settlement pattern as superior to all others both in the interests of the Jewish society and as a social ideal, often placed the representatives of the moshavim in a defensive position. They were compelled to fight for recognition as legitimate realizers of the ideal of the workers' movement and for an equitable allocation of resources.

Because of their more cautious and less dynamic approach, the moshav representatives remained a minority in the center and often resisted the predominant stream of thought. They regarded themselves as representing the working farmer as opposed to the conquering settler of the kibbutzim.

> We were aware that we were bringing with us our own special qualities necessary for the upbuilding of the country.
>
> The readiness to work relentlessly, the organic sense of belonging shared by all members of the family, the constant preoccupation with the matters of the farm . . . however, our participation in the Agricultural Center, both in discussions and in management of its affairs, brought about curious results: it was as if a tractor and an ox were put to plow together, and the match had not worked out well.[19]

To sum up, in their general attitude to the problems of settlement, manifested in their requests for more planning, more consideration for the needs of the farmer, larger basic budgets, and the acquisition of land in the central area by the Keren Kayemeth, the moshavim were not very successful, for the majority refused to let routine practical considerations affect the settlement vision. By the same token, they also got a smaller share of resources, pure and simple; there were thus areas in which they, the "base-born" intruders, were not, in practice, allowed to settle, the land being earmarked as a collective preserve.

So much for relations in the agricultural sector proper. A few words must now be said on the moshav movement, the Histadruth, and the main political parties. It will be recalled that at the end of the first period (1927), an attempt was made by the General Federation of Labor, the majority of whose members were hired workers in the towns, to enforce its authority on the workers' settlements, not only in the political-ideological sense, but also in the legal and economic spheres. (This was the result of the trend towards centralization then predominant in the Histadruth.) The practical manifestation of this tendency was the establishment of the Nir Company, affiliated to the Histadruth. This new institution was viewed as a comprehensive body responsible for the direction and supervision of all the cooperative societies to which the moshavim and kibbutzim were affiliated.

The cooperative Nir had two main purposes, one directed towards the settlement authorities, and the other towards the settlers. For the settlement authorities, it was to serve as a legal representative of the settlers and force these authorities to modify their policies in accordance with the wishes of the Histadruth. To carry this out, Nir had to become a real power and acquire economic resources. Consequently, it requested that all the land leased to the settlers by the Keren Kayemeth be centralized under its authority and proposed to lease it to the settlers as secondary tenants. Negotiations with the Keren Kayemeth were to be conducted neither with the individual settlers nor with their self-governing institutions, but with Nir. The same arrangement was to apply to the farm itself: Nir would receive the credits given to the settlers by the settlement authorities and transfer them to the individual settlers, and when the latter had redeemed them, Nir would become the legal owner of the farm.

In the internal sphere, the constitution of Nir contained a number of clauses guaranteeing the control of the collective over the individual in the sphere of farm planning. It also included clauses guaranteeing the payment of taxes to the moshav and granted the moshav the authority to expel a settler member if 75 per cent of the members voted for his expulsion. Moreover, in order to assure Nir's control over the governing bodies of the moshav, there was a clause granting the representative of Nir the right to attend all general meetings of the moshav and giving him the right of veto.

This desire to enforce Histadruth control on all spheres of the settler's life met with fierce opposition. The leaders of the veteran moshavim opposed centralistic control on principle, even if it was Histadruth control. They regarded the constitution of the new organ as a negation of the freedom of the individual and as a transformation of free people into a "herd." Discussions concerning the establishment of the new institution continued into the thirties; opponents appeared in all the Histadruth circles and even outside the workers' settlement movement. Thus, while the new body was set up, it did not actually become an influential factor, and the clauses of its constitution remained on paper. Matters such as farm planning, progressive taxation, and the maintenance of the authority of the moshav institutions over the individual settlers, which, according to the Nir constitution, were to be under its jurisdiction, did come under the care of the moshav movement. There was, however, a basic difference: these clauses were not implemented as an order from above, but were accepted democratically in the conferences of the movement.

Politically, the moshav movement was affiliated to the largest party in the Histadruth—Mapai, founded in 1930. The majority of the moshav settlers were also individual members of this party, although they were under no obligation to join, and some in fact opted out. None of the moshav members owed allegiance to the more radical socialist parties of the Histadruth—Hashomer Hatzair and Poalei Tzion. It will be recalled that two of the five kibbutz movements—Hakibbutz Hameuchad and Hever Hakvutzot—also belonged to Mapai.

At the beginning, the representation of the moshav movement in the party center was equal to that of Hakibbutz Hameuchad, each of them having in 1930 one delegate in the twenty-member center. In the course of time, the representation of the kibbutz movement increased proportionately more than did the representation of the moshavim. In 1938 the moshavim had six and Hakibbutz Hameuchad eleven delegates in the center, which now numbered forty-seven members. At that time, the number of moshav members belonging to the party was almost the same as those of Hakibbutz Hameuchad. This demonstrates the rise in stature of Hakibbutz Hameuchad and the increase of its influence in the political arena, and thus also in the determination of Histadruth policy and general settlement issues. In contrast, the weight of the moshavim movement was very small, and it lagged far behind the kibbutz movement in ideological and political influence. This can easily be seen in the small number of delegates of the moshavim in the central institutions: in 1933 there were no moshav delegates on the central committee of the Histadruth; in 1939 they had two out of a total of thirty; and in 1942, three out of fifty-two. Hakibbutz Hameuchad, on the other hand, which was not represented at all on the 1930 committee, increased its representation to three in 1939 and to ten in 1942.

On the whole, the representatives of the moshav movement also showed less interest in general questions not directly concerning settlement than did the kibbutz delegates; they were less involved in broad issues and in the juggling for power. In a certain respect, their position was thus one of loyalty to the official leadership of the party, and they usually followed its cues. This obedience, however, was not rewarded in the one area vital to them: settlement. The party continued to be largely swayed by the kibbutz viewpoint and interest. This was so not only because the Mapai collectives constituted a stronger pressure group within the party: in the Histadruth as a whole, *all* the collective movements, including those not in Mapai, formed a united front, preferring considerations of self-interest to loyalty towards the representatives of a different settlement pattern of the same party. In consequence, the moshav stood alone against a coalition, as it did in the Agricultural Center. This was a great practical disadvantage, since the political delegates of various parties to congresses and institutions of the Zionist movement formed a direct channel of influence, and a pipeline for mobilization of resources; the moshav's inferiority meant that here, too, its interests were not properly represented. This was the source of the moshav movement's demand (1937) to present a separate political list which would guarantee them a proper hearing and not to rely on representation within Mapai. Party pressure defeated this attempt at semiautonomy, and the thirties became a time of increased bitterness and recrimination toward the party for a lack of appreciation for the moshav's achievements, demands, and proper rights.

In the late thirties there was a change, at least on the ideological level. The social reality in the Yishuv had been changing since the beginning of the Fourth Aliya, and Mapai was gradually becoming a nondoctrinaire, popular workers' party, which included and accommodated many groups, opinions, and interests. It now contained urban workers, who could hardly be considered as pioneers of settlement at all; and the condemnation of the moshav as a half measure, which had characterized Achdut Haavoda of earlier times, could hardly be maintained. However, this blunting of the ideological edge did little but provide a symbolic sop; in the corridors of power, the moshav remained for the whole period a lame also-ran.

It can thus be seen that the organizational inferiority of the moshav movement to the collective ones had its sources in a variety of interrelated factors. A basic disadvantage was the lack of militant social reformism, reflected so often in the preceding pages. In the war of groups and ideas, the advantage seems always to lie with the enthusiast, the extremist, and the missionary. Closely related was the difference in the two respective images and in what they seemed to offer to society as a whole; on the one hand, plodding, planning, spending and personal security; on the other, the larger vision of conquest, pioneering, and sacrifice. In a context of rapid development and of nation-building, the verdict was on the side of the latter. Then, the agrarian and social system of the moshav was, of course, a much weaker framework for mobilization of manpower, resources, and support for a supracommunity organization. First, as we have seen, its voluntary and individualistic basis prevented the articulation of an over-all authority structure, which could command the complete obedience of villages and settlers alike; and secondly, it could free and provide much less public personnel for positions in and outside the movement.

Perhaps of the greatest importance in this connection was the different social and community basis of the two movements, which was guided and selective in one, and spontaneous and free in the other. Initially, the differences had been small; and Kfar Yehezkiel—to give but one example—came from the stock as well as the party of Ein Harod. The moshav, however, had neither the emphasis nor the institutional mechanisms for sustained recruitment and indoctrination of manpower reserves that could continue the original commitment. In this way, the nucleus of the founding fathers became a minority in a popular movement, in which most members were ideologically undedicated and politically apathetic, and many villages socially insecure and organizationally weak; a movement, moreover, that by its very nature required constant attention and energy for its internal problems, rather than for relations vis-à-vis the centers of power.

Of course, this organizational weakness in a sense reflected the moshav's particular strength and qualities, and it emphasized, in fact, the success of the individual village and of the pattern as a whole under conditions not particularly favorable. At the same time, the fact that the moshav has produced a smaller part, and has certainly eaten less, of, the national settlement cake is here demonstrated and explained. The following period, to which we must now briefly turn, represents indeed a successful attempt to redress the balance.

The Third Period—Population Explosion (1948–   )

The third period in the history of the moshavim movement is acted out in a strikingly different setting, that of new statehood. We cannot, of course, examine the many changes involved within the present context, especially as the period of independence is marginal to our main theme.

The State of Israel was, immediately upon its establishment, faced with the twofold problem of developing the country and of accommodating and absorbing waves of immigrants. For various economic, security, political, and ideological reasons, rural colonization became a major vehicle of both processes, and a significant part of the newcomers had to be channelled to agriculture and to underdeveloped areas. Since most of the people concerned had no previous agricultural experience, completely independent and individual farming was not considered practicable for their settlement. In order to make introduction to agriculture easier and more gradual, it was necessary to relieve them, at least initially, of the ultimate responsibility for the productive process, as well as to provide special and intensive training facilities.

All these considerations made the best solution a closely knit and supervised economic and community organization, within which the absorptive agencies might operate. This decision was, of course, also influenced by the fact that the various pioneering movements had already previously developed several types of the corporate village, collective or cooperative; and these well-tested, ideologically sanctioned, and politically supported forms could now provide fundamental colonizing models, organizational know-how, and experienced personnel.

The choice thus lay, basically, among the main forms developed within the workers' movement; and of these alternatives, the moshav pattern was considered the most flexible, adaptable, and undemanding, and thus more suitable than any type of collective, or even a moshav shitufi, for the settlement of non-selective and socially and culturally heterogeneous groups. Moreover, mass immigration and settlement was the one challenge and duty which the collective movements found themselves unable to accept, estimating—and it seems rightly—that none of them could sustain, without disintegrating and losing their essence, ideologically uncommitted and amorphous groups unsocialized into the kibbutz way of life. It was thus the hour of the moshav, and the movement recognized it as such: on the one hand, it would realize a national function that its founder thought it most conveniently suited, absorption of immigrants,[17] and thus redress the long ideological and political dominance of the collective; and on the other, it could attempt to regain ground previously lost in the movement itself and recreate the moshav in its purer form.

The late prime minister L. Eshkol (then head of the Land Settlement Department) during the Eighteenth Agricultural Convention in 1949, acknowledged:

> The great stream of popular immigration brings with it a different kind of man from the one we have become used to over tens of

years, years of breakthrough of the pioneering movement; a kind of man who, due to circumstances, lacks perhaps the old social qualities, but who may have other qualities to enable him to adjust to our reality. Be that as it may, this is a fact. And over the next five or six years, they will come to the country, and will have to carry on, and on them we have to count. We therefore have to struggle and to strive to turn the new immigrant as he is into a farmer, a creator of agricultural villages embodying the new way of life.

Or A. Assaf, then secretary of the moshav movement, in his book on moshavim (1950):

Now especially, with the establishment of the State and the mass immigrant settlement—settlement of people who colonize the land as a means of making a living without having been ideologically prepared for it and without having profited by any rural experience in Israel, and whose ideas of work and property and their enjoyment are copied from other lands and other systems—now especially, it is necessary, even imperative, to consolidate our way of life and protect it, so as to prevent deviance and to lead us on the path we have chosen, the vitality and importance of which are now more than ever crucial.[18]

The enormous problems and the possibility of failure of such an undertaking were, of course, realized; and the movement, together with the Land Settlement Department, did not ignore the difficulties of the size and tempo of the immigration, and the qualities of the immigrants. However, they also had some grounds for a spirit of guarded optimism: most of the objective hardships which had confronted the earlier settlers could now be eliminated. Before the establishment of the State, it will be recalled, the prospective farmers had to wait up to eight years before land could be allocated to them. Similar difficulties also had attended the provision of capital, and the institutions of the Jewish community had only a limited ability to plan for and modify the market situation. As a result, the full development of the farm had to be deferred for a long time, with settlers obliged to resort to outside work; agriculture as a whole faced a difficult competition with the much more primitive, but much cheaper,

Arab production and labor.[19] Now, however, the greater resources of the State could smooth the path of the settler and compensate for the difference in the human material and in the size of the absorption problem. This was so particularly in two respects: the settling authority could pay for a local team of extension workers in each new village of about seventy households; and, it was able to work according to an intensive development and capitalization schedule, leading up to a potential farm consolidation within seven years.[20]

No less impressive was the parallel commitment of the movement itself: it would put at the disposal of the new villages all the organs and institutions of the Center and open the movement framework and membership to all; it would mobilize from among the veteran moshavim volunteers to settle in the new villages and to provide, at least during the first years, the person-

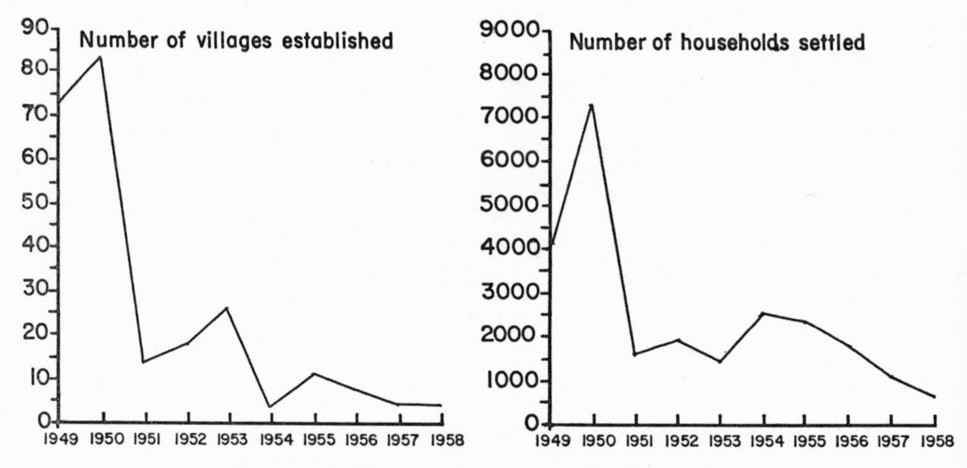

Chart 4. The new moshavim in the years 1943–1958

nel for the instructor team;[21] and it would, last but not least, provide for a greater flexibility and allow a longer time for the actual realization of a full-fledged moshav pattern by the new members.

The realization of this undertaking began almost an once.

Table 23. The new moshavin, 1949–1958

Moshavim	1949	1950	1951	1952	1953	1954	1955	1956	1957	1958	Total
Villages established	73	84	14	19	27	4	12	8	5	5	251
Households settled	4,086	7,265	1,609	1,942	1,457	2,487	2,340	1,818	1,118	645	24,767

Note: Households settled include also those distributed among previously established new immigrant villages.
Source: D. Weintraub, unpublished data.

Tables 23 and 24 and Chart 4 show the growth and the scope of the scheme described, in terms of villages established and households settled on land, and in respect to distribution by farm type and geographical area. Table 25 on the other hand, clearly indicate the social complexity of the program, reflected in the extreme heterogeneity of the villages. (The data relate to the year 1958–59; however, since by then settlement had become a dribble, the figures give a good idea of the situation as it is today as well.)

The diversity of areas of origin shows the sheer magnitude and challenge of the task undertaken. It is no wonder, therefore, that the settlement program soon began to run into trouble. First, there was a large number of leavers; over a period of ten years, 8,587 households (or 34.4 per cent) of those settled left the moshav altogether (excluding transfers from village to vil-

Table 24. Distribution of farm types and areas of settlement in new villages, 1958–1959

Area	No.	Farm-type	No.
Northern	53	Field	57
Central	72	Mountain (mixed)	38
Mountain Jerusalem			
sub-district	33	Hills (mixed)	23
Galilee sub-district	29	Dairy (mixed)	121
Lachish	30	Citrus	12
Southern	34		

Source: D. Weintraub, unpublished data.

Table 25. Structure of actual settler population in new immigrant moshavim, 1958

Area of origin	Country of origin	No. households	% households	No. villages
Eastern and Central Europe	Poland	867	5.3	76
	Hungary	1,363	8.4	
	Rumania	2,183	13.5	
	Czechoslovakia	418	2.6	
America				2
	U.S.A.	48	0.3	
	Argentina	88	0.5	
The Balkans				10
	Yugoslavia	250	1.5	
	Bulgaria	350	2.2	
	Greece	130	0.8	
North Africa				85
	Morocco	2,454	15.3	
	Algeria	31	0.2	
	Tunisia	1,026	6.3	
	Libya and Tripolitania	1,113	6.9	
Middle and Far East				71
	Turkey	158	1.0	
	Syria and Lebanon	75	0.5	
	Egypt	312	1.9	
	Yemen	2,168	13.5	
	Kurdistan (Iraq)	1,500	9.3	
	Persia (other)	786	4.8	
	Cochin	246	1.5	
	China	38	0.2	
Others		575	3.5	7
Total		16,179	100.0	251

Source: D. Weintraub, unpublished data.

lage). Table 26 summarizes this population movement. For comparison's sake, the leavers are combined together with the totals of settled (as in Table 23).

Secondly, in a few of the new moshavim development was not following the prescribed schedule and achieving the planned effect, with crisis characterizing both the individual and the village level. There was, thus, a growing trend towards part-time farming, with a consequent waste of allocated resources. These

Table 26. Out-movement from the moshav, 1949–1958

Years	Left		Remained		Settled	
	No.	%	No.	%	No.	%
1949–50	5,215	45.9	6,136	54.1	11,351	100
1951–52	884	24.3	2,687	75.7	3,571	100
1953–54	738	18.7	3,206	81.3	3,944	100
1955	707	30.2	1,633	69.8	2,340	100
1956	327	18.0	1,491	82.0	1,818	100
1957	328	26.0	890	74.0	1,218	100
1958	388	55.3	357	44.7	745	100
Total	8,587	34.4	16,400	65.6	24,987	100

Source: D. Weintraub, unpublished data.

were gross deviations and inequalities in the planned farm production and income. There also occurred cases of social disintegration of communities that were incapable of functioning as cooperative and municipal frameworks.

No doubt, some of the phenomena observed could immediately be traced to self-evident economic factors. For example, not a few of the villages (particularly in the mountain regions), which had been established primarily on other than purely economic considerations, had insufficient land and water from the beginning. In many cases, too, villages were founded without any detailed planning or preliminary survey of the soil, and this sometimes resulted in the establishment of settlements without a sound economic basis. More generally, the supply of capital to settlement (which depended largely on Jewish donations abroad) fluctuated and often fell short of over-optimistic estimates. Similarly, agricultural planning had been largely at the farm level and not based on market requirements, in consequence of which there were inevitable surpluses and price crises.[22]

Administrative difficulties and snarls also hampered various activities at the village level and higher up. Lack of adequate administrative routine, organs, and personnel thus caused snags in marketing, in credit arrangements, in supplies, and in many other areas.[23]

These factors, however, explained only part of the situation, which could be traced chiefly to incompatibilities between the moshav pattern and the social characteristics and aspirations of the different settler groups. The nature of these incompatibilities cannot be documented and analyzed here. It forms, in fact, the subject of another volume.[24] They showed, however, that the institutional framework of the moshav was straining dangerously under the impact of problems and trends that had been unknown or rudimentary in the "citrus-thirties," or among the most emancipated second generation. The moshav pattern was in fact finding it difficult to socialize and absorb mass settlement on a voluntary basis with the existing institutional, social, and economic controls; and the original idea was thus in a danger of being swamped and subverted. It is in these terms that the remedy of legal safeguards and limitations, incorporated in the Moshav Bill, must largely be understood—a Draconian law to meet harsh reality.

As we saw above, the measure has disturbed the movement deeply, with lines of division running across veteran as well as new groups. The opponents emphasize ideological and moral issues and predict harmful effects to enterprise, rational development, and self-expression. The supporters, including the Center hierarchy, stress danger of disintegration and of loss of villages to agriculture and to the moshav and—at least implicitly—fear the possible loss of power and position, so painfully won by the movement.

PART III

CONCLUSIONS

8

Comparative Analysis

The previous chapters have offered the reader, first, the history of the development of three villages, each a prototype of a style of settlement developed in the Jewish community in Palestine (and to some extent also in Israel); secondly, in conjunction with this, we have described the three central settlement organizations that supported these prototypes.

The above presentation has, we hope, illustrated the areas of similarity and difference that exist between the three forms of settlement and their organizations. The purpose of this chapter, however, is to present these areas in a more systematic and analytical way, using as reference points the problems raised in the introductory chapter.

Settlement Patterns

It is not our intention to offer any final evaluation of the successes gained, absolutely or relatively, by the various types of settlement studied in this report. We are concerned, rather, to trace the ideological, agrarian, and social configurations presented by these types, and to see their potential in respect to internal development, absorption of change, and introduction of innovation. The analysis falls under three main headings: values and normative predispositions, and their development over the years; the agrarian system and its implications; and social and political structure.

VALUES AND NORMATIVE PREDISPOSITIONS

None of the settlers of Petach Tikva, Ein Harod, or Nahalal were born at these places. With the exception of a few special cases, they and their forefathers were distant from both the mental outlook and social and economic problems of an agricultural way of life. None of them were forced to turn to village life, and all had alternative opportunities. The decision to take up this way of life was purely *personal;* the frameworks that were established remained voluntary, with regard to the right to join or to leave, and the limitations that the settlements imposed on themselves did not contradict this essentially voluntary nature. A systematic recapitulation of the settlers' normative predispositions and motivations [1] is therefore especially important here.

A basic distinction at once comes to mind between the motives that led to the establishment of Petach Tikva and Ein Harod. The founders of the former group sought primarily the renewal of the traditional pattern or a new approach to the worship of God and the keeping of the religious commandments,[2] while the settlers of Ein Harod were determined to develop an almost complete alternative to traditional Jewish society.[3] The first group thus emphasized their identification with, and their desire to continue, their original society, with the exception of certain customs that were to be revised; the settlers at Ein Harod, on the other hand, deliberately intended to "inherit" the place of the traditional society by eliminating most of its fundamental characteristics.

Because of this contrast between the two groups, farming served very different needs in each of them and was approached from different points of view. Thus, in Ein Harod agricultural labor became the vision of a complete way of life. This image combined, as we saw, personal and national pioneering and regeneration, collective production, and a communal society. And the kibbutz, seeing itself as a fundamentally new and revolutionary pattern of life, also developed an orientation towards expansion and prepared to determine as well as carry out national objectives.

The missionary outlook, characterized by a high degree of militancy, was completely foreign to Petach Tikva, whose emphasis was turned inward upon itself in a narrow, particularistic way. Even there, the approach to the problems of settlement was largely institutional or extrinsic, and it was not conceived of in terms of individual and community values. In this way farming, though symbolically tinged, was mainly a means to an end and essentially an occupation. Similarly, while the idea of cooperation was not unknown in the moshava, it was approached in a pragmatic way.

In Nahalal, the sense of creating an alternative pattern was less evident than in Ein Harod; while it may have existed upon the arrival of the group in the country, it weakened with the passage of time. The settlers were not ready confidently to proclaim themselves the sole heirs of the older society and accepted the legitimacy of other ways, which they freely borrowed—or rather recreated. From Petach Tikva they took family farming and turned it into a family-farm culture, while they followed the collective ideology in respect to pioneering, manual and self-labor, and simplicity of life. This over-all social vision, however, was in the moshav inevitably tempered by family individualism and its more diluted revolutionary zeal: it thus put less stress on the proletarian class aspects of agricultural labor, which were strong among the radical socialists of Ein Harod; and it substituted intensive cooperation and equality of life chances for collectivism and egalitarianism.

These, stated briefly, were the different ideologies of departure of the three prototype patterns. In no case, however, were the original ideals preserved in their pristine state, and the extent of change in them—especially the attitudes adopted to such change—were of considerable significance.

As regards the strength of change, priority is undoubtedly to be given to Petach Tikva. Shortly after the foundation of Petach Tikva, a division arose between the settlers and the Jewish community of Jerusalem from which they had come. The focus of the organizational and ideological contacts moved from the orthodox elements of the Old Yishuv to the Jewish national movement located in Eastern Europe. This was not to be the final stop, as evidenced by their links with Baron Rothschild, J.C.A., and at a later stage by the violent conflicts with the center of political power in the Jewish sector of the country, during the period of the British mandate. Even the shift in their self-image from that of independent farmers to economic and administrative dependents, during the period of the Baron, was sharp, if not smooth.

It is interesting to speculate what prevented the stagnation, or even the disintegration, of Petach Tikva during the period of the Baron's patronage. The answer is doubtless complex. It can be argued that the period was too short to irreparably damage the settlement; it can also be claimed that the Baron's investments (especially the introduction of the cash crop) and those of J.C.A. did eventually bear fruit. The geographic location may also be significant, especially the proximity to centers of overseas purchasing and marketing. All these probably played a part, but it seems to us that the ability to overcome difficulties and the degree of flexibility exhibited point to another factor: the absence of any necessary and normative a priori connection between the ideal social structure envisaged and the type of agricultural and economic activity. The lack of such a link preserved traditional social structure, especially the communal organization and cultural and spiritual characteristics. It also facilitated the radical transformation of those motivational values that drove towards agricultural activity, and determined the actual nature and content of these activities.

This accounts for the fact that by the end of the second period, less than ten years after the founding of the settlement, agricultural labor had ceased to be seen as a goal in itself, to be pursued regardless of economic considerations. The instrumental attitude to land and to agricultural labor, which had doubtlessly been latent even earlier, now became the dominant feature, although its results did not become evident until the third period, when the Baron's administration came to an end.

It must be stressed that the new economic initiative displayed by the farmers of Petach Tikva had no such religious legitimation as had been given a few years previously to agriculture as such. The new initiative was merely a successful response to new desires for higher standards of living and economic consolidation. This led to a willingness to abandon the existing subsistence economy and to turn to a market-oriented economy with all that this implied. Thus, the change was radical and the lack of loyalty to certain basic values was clearly evident. But the settlers had no guilty feelings; the value transformation in this sphere received full acceptance. The changes were facilitated also by the slow and undramatic rate of social change.

To the extent that changes in social and cultural characteristics did occur, these were the result of external pressures, and they remained, to a marked extent, both non-institutionalized and non-regulated. We refer especially to the process of demographic and economic growth that introduced new elements into the colony, especially craftsmen and an agricultural proletariat. It was this confrontation with new elements that led to an attempt to renew the ties between the pattern of economic activity and the social character of the village that had to be defended, especially because of the presence of hired Jewish laborers. The demand that Jewish labor be employed exclusively was felt by the farmers to constitute an economic threat. This was coupled with the fear that if these laborers were to gain full civic rights in the settlement—and this was implied in the demand for Jewish labor—then the specific traditional-religious nature of the village would be destroyed.

We have previously said that Petach Tikva holds the monopoly of radical change, accompanied by a *lack* of negative or ambivalent attitudes, with regard to such changes. On the other hand, Ein Harod preserved its basic values, almost in their entirety, until the 1940's at least. We find a continuing militant-socialist orientation, and the collectivist pattern in production and consumption, the egalitarian division of labor and its rewards, and the great willingness to accept new members were all maintained. The same applies to the emphasis placed on the salience (though not on the exclusiveness) of mixed farming and on the optimum of self-labor.

It would be incorrect to claim that there had been no deviations or doubts as to the feasibility and institutionalization of these principles on the part of the kibbutz. However, in most cases the tendency was to regulate these changes and deviations by making a more liberal and permissive interpretation of the abstract principle in question. Thus, individuals were offered a greater range of choice in the field of personal consumption (clothing, furniture, the use of free time, etc.) and elements of social life were transferred from the collective to smaller subgroups, especially the family. We find, then, a change from a mechanical egalitarianism to a state of substantive equality that attempts to realize the principle of "from each according to his capacity, to each according to his need." The major deviations developed in connection with the principle of self-labor and the opposition to hired labor. This problem had not existed during the stages of slow economic growth, but when the settlement grew and the shortage of both seasonal and skilled labor was felt, the matter became urgent.

The long period of economic weakness made Ein Harod anxious to encourage development when it became feasible; but this led to an increased need for hired labor. This was, however, always regarded as deviant in nature and never received legitimation. Solutions were sought which would both ensure the continued growth of the kibbutz while preserving the basic norms of self-labor and the non-exploitation of the work of others.

Similar but not identical developments occurred at Nahalal. A central impetus for the foundation of the moshav lay in the wish not to be dependent on the market as *consumers* and in a reluctant orientation to the market as *producers*. This was accompanied by a rejection of a mechanical egalitarianism and of the collectivistic structure: support was given to the ideals of cooperation, but the family held the central position as the unit of production and of consumption.

The ideal was of a simple life, with an emphasis on austerity and a rejection of the cult of property and the pursuit of luxuries. Above all, there was a complete denial of the concept of farming as merely an occupation, and emphasis was laid on the wider considerations of agriculture as a morally laudable and productive way of life, a complete antithesis to urban life. As with Ein Harod, there was no *manifest* transformation of values at Nahalal during most of the period under consideration, until the late forties. The official ideology remained in force, and there was no reluctance to enforce sanctions against deviation from the norms, especially with regard to hired labor. However, later these deviations become more extensive, and the regulative power of the moshav committee diminished.

Serious and honest attempts were made to institutionalize the changes and to define more permissively the restrictions regarding hired labor and the supremacy of mixed farming. However, it is doubtful whether these efforts were sufficiently strong to restrain the momentum of the external and internal pressures that gave rise to the changes. This was especially so as regards the general rise in the standard of living and the emergence of a more selective consumers' market: demands became more varied, and intense, so pressure was exercised towards specialization and the elimination of unprofitable branches. Thus, the tendency has been to move from the ideal of autarchy to the type of farm completely integrated with the Israeli market system, with all that this implies. Although the changes appeared, as we have seen, late on the scene, many members of Nahalal are already critical of the authority of the community not only to regulate production and choose market targets, but also to limit the size and the expansion of the unit of production and the standard of life of the individual members.

As was mentioned above, many concrete solutions introducing secondary institutionalization were adopted. However, it is difficult not to conclude that the moshav met with greater difficulties than the kibbutz in dealing with the continually increasing discrepancy between the normative predispositions and motivations and the trend towards economic development. These discrepancies were, in part, due to the objective situation, in which economic growth was greater than that which could be supported by the labor force of the productive unit (the family), and where there was a need to absorb immigrants. At the same time, the weakest point seems to be the desire, conscious or unconscious, for economic autarchy and a partial physical and mental segregation from modern market processes. There is no doubt that a process of liberation from this mental outlook is at its peak now—especially among the second generation—but it is difficult to speak, at least during the 1940's and 1950's, of a transformation in values similar to that which occurred at Petach Tikva after a very short period.

On the other hand, Ein Harod was free of these uncertainties. The conception of the kibbutz movement as an alternative social pattern prevented an apathetic attitude towards the problems of agriculture and its position in the productive processes of the general society. There was a definite aim to revise the capitalist market structure, but this did not include any desire for economic withdrawal and segregation. There was, rather, a continual desire to specialize and to be integrated as producers competing within the general market processes. In this respect the entrepreneurial drive of the kibbutz resembled that of the settlers of Petach Tikva, also free of such misgivings. Both allowed themselves to participate in the wave of economic expansion by the resourceful utilization of opportunities, and they did not suffer from any ideological uncertainty.

To sum up, the settlers of Petach Tikva and the founders of Ein Harod and Nahalal were all innovators. The first were actually pioneers in the choice of agricultural work as an occupation and a way of life. But the initial drives of the first settlers of Petach Tikva were not connected with a revolutionary ideological transformation; the aim was rather to revive and renew their religious observances while rejecting the unproductive way of life of the Jews of Old Jerusalem, who lived on charity. This was the source of the limited aims of the founders and their pragmatic approach to economic activities, accepting them as long as these met the religious requirements. The experience of Petach Tikva and other farming settlements made up of religiously observant members serves to prove that the potential contradictions between a rational approach to farming and economic initiative on the one hand, and religious commandments on the other, are not so extensive and can be resolved within traditional frameworks.

By contrast, the initial impetus operating among the settlers of Ein Harod was tied to an absolute and all-inclusive ideological transformation and had more extreme final goals. This drive was diluted and circumscribed to some extent in Nahalal, but here too the changes planned were of a completely different nature than those conceived by the innovators among the settlers of Petach Tikva.

The Agrarian System

The agrarian systems of the three types of settlement, as initially established and modified in response to challenges of reality, reflect essentially the value predispositions of the settlers and the changes in these values over time. It is, indeed, in these terms that the striking differences between the three patterns must be understood.

A basic difference between Petach Tikva on the one hand, and Ein Harod and Nahalal on the other, lay thus first of all in *ownership and tenure* of land. In the moshava, it is a private freehold, with no institutional restrictions on sale, distribution, and inheritance. The lands of the kibbutz and the moshav are nationally owned, any sale or alienation being publicly controlled. In both cases, though, tenure is secure (forty-nine years and renewable), inter-generational continuity being assured in the first by the corporate tenure itself, and in the second by a right of inheritance. In the moshav, however, security of tenure is limited as regards transfer, both to other holders and to more than one heir; and these restrictions guard against both family farm fragmentation and concentration of land in larger units.

Considerable differences exist also in the crucial sphere of *land use*. In Petach Tikva, the family started as the farm unit, both in management and entrepreneurship, and in actual cultivation or labor. However, while the first has remained unchanged—freedom of decision being here, in fact, much greater than in Nahalal—cultivation has become mixed or combined, hired hands being used legitimately according to economic considerations.

Cooperative organization between units of production was here also an early and transitional episode. Since then, arrangements are determined by objective conditions. In other words, cooperation represents an attempt at maximum utilization of manpower resources and is not even marginally a matter of ideological principle. With regard to financing, preference had originally been for own resources. However, following a philanthropic interlude, commercial loans for specific purposes, together with reinvestment of earnings, have become the accepted pattern. This has meant an emphasis on savings and capital growth, especially through cheap labor and control of marketing.

Non-economic value considerations in the branch structure also disappeared early in the colony. In the beginning, Petach Tikva too established branches of production on the basis of symbolic significance rather than practical value. Indeed, mixed farming was to be implemented here not only because of considerations of stability and flexibility in the seasonal allocation of labor, but also in order to achieve the ideological aim of autarchy. This, however, was but an episode at Petach Tikva. The farmers rapidly developed new branches, and vineyards designed for export were established for intensive cultivation. This was the beginning of intensive capital agriculture and was followed by unsuccessful attempts to grow almonds, before the citrus plantations that were to bring economic prosperity were developed.

In Nahalal, too, the family is the unit of cultivation. Here, however, the terms of reference of this unit are markedly different: on the one hand, its decision-making and autonomy are circumscribed much more by public planning and by cooperation; and on the other, insistence on family labor is complete. These last, together with the above mentioned regulation of land transfer and inheritance, are two mainstays of the classical family-farm pattern, where the holding is the center of activity and identification. Another mainstay is, of course, the mixed farm, which has consistently been maintained. This is so even though, towards the close of the period discussed, tendencies towards greater selectivity and specialization develop; the dominant trend remains a balance between branches. Indeed, Nahalal has continued to uphold, to a considerable extent, the practice of autarchy on both the consuming and the producing level. On the other hand, compulsory cooperation binds the different units into a comprehensive, multi-purpose framework of production, supply, marketing, credit, and other services.

Ein Harod is similar to Nahalal in its emphasis and reliance upon public capital and subsidies. While both have had recourse to private sources, it appears that quantitatively these performed only a marginal role. Likewise, the use of self-labor (although, of course, not family labor) has played a dominant role, as have field crops and the dairy herds, resulting, in practice, in an overcommitment to mixed farming. However, here these considerations were given much wider definition, so that diversified crafts and outside work for wages were included, in contrast to Nahalal's narrower concept of purely agricultural undertakings. Thus, while Ein Harod, unlike Petach Tikva, was restricted in its relations to the money market, it was freer than Nahalal in its dealings with the labor market and in producing and marketing goods.

However, the fundamental distinction of Ein Harod, which sets it apart from the two other villages, is the nature of its unit of production. This unit, in management and entrepreneurship as well as cultivation, is collective, as are all economic and farm services, distinct from the pragmatic-contractual arrangements in Petach Tikva and the compulsory cooperative relations in Nahalal.

Collective production and management undoubtedly mean that Ein Harod can enjoy the economics of size, with much greater scope for both volume and differentiation. The emphasis on mixed farming, weaker here than in Nahalal, thus can co-exist in practice with high-level specialization. In this respect, as in others, *production* in the kibbutz is less circumscribed by non-economic restrictions, and it is managed in many respects like a large capitalist enterprise.

This feature is also reflected in the division of labor between the sexes. Thus, Petach Tikva maintained a clear-cut allocation, with the men engaged in or supervising field work, while the women were responsible for the household and home gardens. Nahalal approached this pattern, with its acceptance of the family as the basic unit of production and consumption. The situation at Ein Harod was different. There was no separation between the functions of consumption and production performed by the family and those services supplied by central institutions. Since the family was not the basic unit of production or consumption, there was no need to make any division along lines of sex, and in the early period the women themselves were most vocal in their opposition to any such role allocation.[4] With the passage of time, however, such a division of labor did arise, with most women engaged in services and the men in the productive branches.

Of all the agrarian aspects, physical structure is certainly the most similar in the three types of settlements, which are predominantly gathered villages. This is so since they have always faced problems of defense, and because of their essentially *Gemeinschaft* community image.[5] However, it is on Nahalal in particular that this layout imposes special limitations: here the wish to safeguard the potential for rational cultivation and mechanization, together with equality of land allocation, result in considerable dispersion of individual plots. At the same time, we have seen that this problem is capable of at least partial institutional solution through cooperative cultivation and re-parcellation, which also increase the actual unit of production and its scope for efficient utilization and specialization.

Much more difficult is the evaluation of the specific problems and the relative advantages of these structures. At the same time, three areas of comparison seem to be most fruitful in this respect.

1. *The nature and scope of the goals which the agrarian system is designed to achieve or facilitate for the farmer and for society.* In other words, what is the relationship between the agrarian structure and the social and the ideological spheres. In Petach Tikva, these aims are essentially economic, even capitalistic, and consist chiefly of stimulating production and profit.[6] In both Ein Harod and Nahalal, the agrarian system serves and symbolizes other goals, embodying a way of life. It is, of course, difficult to evaluate, except in reference to social values and individual aspirations, whether a narrow economic, or a broad social-ideological, agrarian frame of reference is better. Inherently, however, the two patterns have specific advantages and disadvantages *functionally:* thus, economic primacy is inevitably freer of obstacles to rational mobilization and utilization of resources and permits greater mobility, both within the rural sector, and outside. On the other hand, a broader interlinking of agrarian with social elements means a presence of checks, balances, and sanctions against "rampant" economic development, in the sense of activities outside the accepted basic objectives; Nahalal and Ein Harod were able to prevent, even at their most permissive, the dramatic agrarian changes, if not destruction, which occurred in Petach Tikva.

2. *The ability of the agrarian system to achieve its aims.* As we saw above, the pattern of Petach Tikva, that of a *private agricultural firm,* is institutionally freer and potentially more flexible than the *collective pattern* and that of the *family farm.* Theoretically, therefore, conditions being equal (which they were not) the agrarian systems developed in the moshav and the kibbutz could have obstructed the path of rapid economic growth and consolidation. Indeed, it was Petach Tikva, free of ideological considerations, which efficiently utilized the cheap manpower resources of the Arab population and attained maximum rationalization in marketing techniques, thus taking the shortest way to economic development.

The data at our disposal do not permit even a partial answer to the basic economic question of the *actual* advantage of the various agrarian systems in productivity, even though, theoretically, the private-firm system seems to be institutionally better for the *rational use* of resources. It is probably impossible to recall completely the necessary details and—even if reconstructed—to evaluate from them the differential productivity of resources, since their formation. It would be technically too difficult to measure the productivity per person or family in parallel periods of time; and the dissimilarity between farming branches frustrates comparisons. Also, different resources are differentially productive, and the significance of their productivity varies, in relation to their scarcity.[7] Last but not least, narrow economic comparisons between a *firm* and a *way of life* seem invalid.

At the same time, possible differences notwithstanding, each of the three settlements realized, during the period studied, a high measure of productivity, so the maintenance of its particular system *did not entail any important economic price.* The data seem to say this is no longer true; and there appears to be a significant difference between Nahalal and Ein Harod as regards the *continued* ability of each agrarian system to maintain economic headway. In this respect, we have seen that many of the agrarian arrangements of Nahalal are now becoming obstacles to further development and intensification; from this point of view, the moshav, unlike the kibbutz, is in need of urgent agrarian reform. The latter has emerged as the more flexible of the two agrarian systems, in the range of institutional alternatives to achieve its aims. This is so because the moshav pattern (in its agrarian, but not necessarily social, aspect) is more contradictory, as well as more rigidly and formally defined.

3. *The requirements of the agrarian patterns in resources.* The three settlements also clearly differ in the *investments* necessary for their creation and maintenance. This applies to economic resources, but even more to social and ideological factors. Here the situation described above appears to be reversed, and the collective pattern of cultivation and consumption, while uniquely suited for ecologically severe conditions, emerges as the

most exacting; it requires the most intensive commitment and the greatest degree of social integration and mobilization. Designed by and for élite groups, it is expensive in terms of scarcity of élite resources and incapable of existing viably on any others.

To sum up, the three agrarian structures analyzed have differential drawbacks and advantages when evaluated from various points of view. Of course, as we pointed out at the outset, the much broader function of the kibbutz and the moshav individually and nationally, makes their comparison with the moshava, in narrow terms of economic-agricultural institutions, spurious. With this qualification in mind, however, the differentials stand out clearly. The *private agricultural firm* of Petach Tikva is easiest (or cheapest) to establish, least complicated in achieving its limited goals and also easiest to dissolve as a rural fact. The *small family farm*-cum *cooperative* of Nahalal is in human terms "expensive" to create and maintain, less institutionally geared to sustained development, and most resistant to agrarian disintegration. The *collective* unit is socially most difficult to uphold, more capable than the moshav to sustain development. (It is, as has been said, economically similar to a large capitalist enterprise.) But it is also more likely to lose its agrarian basis.

Social Structure and Internal Government

In our analyses of the three settlements, we did not deal systematically or in detail with concrete social and political events and processes. Not only would such a reconstruction be difficult from mainly documentary material, but we preferred to stress the basic potential of the societies described—in terms of group and institutional structure—to integrate and expand, and to identify with and organize in support of the aims and policies of their élites.

Of the three settlements, it was Ein Harod which set up an almost classical example of a social framework that could be called a mobilization system.[8] This implies an organization where the members are subject to continuous mobilization without the restrictions of any particularistic interests. The entire set-up at

Ein Harod, and all its institutions, was designed with this aim in mind, since this was felt to be the central function and point of superiority of kibbutz society. Thus, the major criterion for the selection and acceptance of new members was their readiness to accept the authority of the collective and their ability to cooperate with its members; this principle made possible the continuous expansion that was the explicit aim of the supporters of the large kvutza ideal. At Ein Harod and other kibbutzim of Hakibbutz Hameuchad Federation, purely economic criteria were rarely adopted in evaluating absorptive capacity; the members of the kibbutz steadfastly fought the central institutions' demands to reduce membership, even in times of grave economic difficulties.

At Petach Tikva, on the other hand, expansion depended purely on the diversification and the development of the colony's economic base. At least as early as the second period, Petach Tikva had two types of citizens. The first were those landholders who had full civic rights and were able to participate fully in public affairs. The second group was made up of hired workers, independent craftsmen, and others who supplied services; they lacked civic rights in practice, and it was only gradually, and over a long period, that they were able to gain these rights and eventually take over the internal control of the settlement.

Nahalal had, upon its foundation, established a rigid upper limit to demographic expansion, with land allocated fully on the existing family basis, including some reserves for natural increase. This was in distinct contrast to Ein Harod and even more to Petach Tikva and implied strictness in the selection of potential members for Nahalal and objective difficulties in the absorption of new settlers.

As to social consolidation, we saw that the founders of the large kvutza ideal were disappointed in the conceptions of the intimate "*Bund*" and *Gemeinschaft* framework formulated by the followers of the small kvutza; at Ein Harod, already in the earliest stages, the social set-up was on the lines of a "sectoral community." Such a community, studied empirically and analyzed by J. Talmon, who is responsible for the term, implies a pluralistic structure made up of a number of both temporary and permanent groupings and of partial segregates within the over-all population, based on geographic proximity, work groups, age and length of residence, and social status. These constellations interact with each other and are very often formed as the need arises to meet the collective's over-all objectives. As could be seen from our data on the development of Ein Harod, this is not a social system devoid of conflicts, and the question of meeting individual requirements, within this collective, still awaits a final solution. However, there is no indication, at least during the period under discussion, that this problem affected the ability to mobilize the population to perform and attain the goals of the collective. This seems to have been true also of other potential areas of conflict that developed at Ein Harod, primarily because of stratification within the population: the growth in social distance between various groups was a priori limited because of the basically egalitarian nature of the social structure.

At Nahalal the primary social grouping was the family. Beyond this, and below the formal cooperative level, we know of sub-groups devoted to common cultivation, ownership of machinery, and similar instrumental areas. There are also several indications of incipient youth culture groupings contrasting with adult culture, especially in the second and third generation. A number of facts point to an independent coalescence into a group by the teachers, administrators, and craftsmen who have only auxiliary farms in the moshav. However, this does not necessarily imply a sharp division between the different groups, and especially not in the field of mutual informal relations; over-all solidarity has in fact been maintained, except for the extreme case of the Moshav Bill (see Chapter 7).

The various families' degree of economic success is a further channel of internal differentiation, especially prominent towards the end of the period surveyed. There is no evidence available to show whether this criterion already acts as a basis for group coalescence; at the same time, the system of progressive taxation and the institutionalization of mutual assistance have not prevented the development of considerable objective differentials. However, much more important than social differences and crystallizations seems to be the conflict and the withdrawal of support now spreading to many basic aims and policies, and particularly those relating to agrarian restrictions and the scope and intensity of social control.

In Petach Tikva the entire economic and social basis for membership was, from the beginning, non-egalitarian, and social and economic polarization was rapid during the second and third periods. The deprived group was not always the same; at one stage it was those farmers who rejected Baron Rothschild's patronage and so lost all political power; at other times, and for a long period, it was hired workers lacking property and having no civic rights in the colony. The antagonism between these laborers and the farmers was magnified by the former's secular, socialist orientations and the latter's religious conservatism. Indeed, Petach Tikva was the main theatre for the conflict between farmers of the colonies and the temporary and permanent workers; the village became a symbol in this respect, and the bitter struggle that went on between the two sides became part of the folklore of the Israeli labor movement. However, total polarization was prevented by the parallel struggle that went on between the younger and the older generation in the village. This centered on the question of modernization of the cultural character of the village and enabled temporary alliances, usually concealed and concerned with specific objectives, to be formed between the youth and the workers. In addition, external dangers from Arabs and foreign rulers produced a fragile framework of solidarity, in spite of the different approaches to national symbols. All in all, Petach Tikva was far from having a unitary mobilization system; it is possible only to refer to various sectors that organized to express specific needs and demands.

The three settlements are also characterized by differences in leadership and authority structure that are directly reflected in their particular abilities to gather up various interests, articulate them, and manipulate the available manpower. All three started with minimal distinctions between the political leadership and rank-and-file members. Within a short time, Petach Tikva became differentiated from the other two. Holders of executive roles (members of the council or committee) were almost completely dependent on *external* élites during the entire first and second periods (the first twenty years of settlement). Initially these external élites were the rabbis of Jerusalem, later the leaders of the European Hovevei-Zion movement, and finally the administrators installed by Baron Rothschild.

A similar situation never developed at Nahalal or Ein Harod. This is not explained by the claim that Petach Tikva was more economically dependent on external factors. Nahalal, and even more so Ein Harod, were no less dependent than Petach Tikva on external elements. Perhaps the reason is that Petach Tikva failed to produce any leaders of national importance during its earlier period, while both Nahalal and Ein Harod were able to produce such leaders. A further possible explanation is that Petach Tikva had, from the beginning, readily accepted a degree of direction and assistance from external leadership. After all, they made no claims to be rebels or to be advocating an all-inclusive social alternative that necessitated reluctant contacts with foreign and external élites. With differing degrees of intensity both Ein Harod and Nahalal evaluated their attitudes to external élites according to the degree of support and assistance they offered to consolidation of the *original* patterns of settlement.

A further point that must be taken into account is that the degree of legitimation given to intervention by élites in the lives of the settlers differed in all three types. This intervention was most intense at Ein Harod. At Petach Tikva, by contrast, social control existed initially, in particular with regard to religious life, but with the passing of time it was actually, if not theoretically, restricted to instrumental affairs, such as taxation and representation before the authorities.

The élite structure, and especially the degree of selectivity, were also different in each case. In Petach Tikva, only landholders were originally eligible for election to the council, while hired laborers and others were deprived, for a long period, of electoral rights. Inequality between landholders also existed: the number of votes allotted depended on the amount of land held. This principle was abandoned, but discrimination and selectivity continued to exist; and if it was less institutionalized, it was no less real. Members of the council were thus for a long time chosen by outside bodies such as the Hovevei Zion organization or Baron Rothschild's staff; and they acted as agents of these bodies, and their role was to coordinate and communicate between the farmers and the external élite.

Ein Harod started its career as a full and direct democracy and eliminated the "living cells" that had introduced a degree of hierarchy in the period of the Labor Battalion. However, the collective and communal nature of the kibbutz soon made the establishment of a varied range of administrative offices, economic and municipal, necessary, and the various voluntary public roles were expanded and institutionalized. Thus, we see the establishment of the large and the small secretariat, and various sub-committees such as those dealing with cultural, health, and membership affairs. The considerable concentration of power in the hands of a small number of members of the secretariat made it essential to develop certain mechanisms of control, such as rotation of roles and close supervision of the elected bodies by the authority of the general assembly of all kibbutz members. In time, and especially with the process of growth and expansion, the need for a defined authority structure was accompanied by the need for specialization in public positions, especially in the economic sphere; considerations of efficient management thus limited the safeguards of job rotation and gave place to the formation of more permanent branch tenures. However, the crystallization of an élite enjoying a preferential position is prevented by the high sense of public service in the kibbutz, by its great social participation, by the lack of formal differential rewards, and by the impossibility of accumulating such rewards.

Of course, the lack of differential rewards is in itself a problem in a context of differential responsibility. However, the same sense of public service assures recruitment to office also under such restrictions. It is, true enough, possible to point to the existence of status groups based on the temporary monopoly of offices and authority over others; but even recently, and certainly during the first twenty to thirty years, their right to exist was based purely on personal esteem awarded to their members because of their loyalty and adherence to the collective goals.

We see that the kibbutz solves most of its potential social and political problems and contradictions by mobilizing high commitment and dedication. However, the considerable extent of ideological involvement in a context of a total framework means that it is intolerant of differences of opinion and incapable of sustaining conflicting approaches. Such differences are aggravated by the open membership policy. Segregating the conflicting groups and breaking off contact between them is socially and physically impossible. There is indeed no place for a great degree of sub-group autonomy and for structures having parallel functions (such as marketing or educational institutions) that can be established so as to guarantee separation. And when the situation becomes tense, there is no solution other than partition of the settlement or departure of the minority. Ein Harod split a number of times during the period surveyed; in 1954 an even more drastic and extensive split occurred because of external political events (the Government's foreign and domestic policies), in the course of which the village was divided into two settlements.

Petach Tikva, on the other hand, was able to grow and develop in spite of continuous conflicts, because the governmental and social structure allowed for an *institutionalized separateness* in the fields of education and welfare services side by side with good neighbor relations and instrumental cooperation on the economic and, to a degree, the political levels.

The governmental pattern of Nahalal was similar to that of Ein Harod, with the exception of certain variations made necessary by the agrarian system and the members' ideology. Initially there was a separation in terms of personnel (in principle, if not always in practice) between the agricultural cooperative and the village council, and only farmers were entitled to vote for the former. The range of authority of both the council and the general assembly was formally more limited than that of the kibbutz's secretariat and general assembly. This was especially so with regard to patterns of consumption and production by the family unit. On the other hand, the authority of the committee of the cooperative was similar to that of the kibbutz secretariat in providing credit, tools, seeds, marketing, and the maintenance of mechanical equipment held in common by all the moshav members. All in all, however, the range of problems was here somewhat different: Nahalal was thus characterized by greater social differentiation and deviation from norms; and since its system of government, chiefly the municipal social sphere, was voluntary and commanded few practical sanctions, it faced serious difficulties of social control. By contrast, the very institutionalization of the family as the unit of production, consumption, and socialization is here a mechanism to contain and isolate potential areas of social and ideological conflict. Since the members of Nahalal are also less involved, as has been seen, in general ideological disputes, the moshav, while usually less cohesive, is in the long run socially more stable.

Settler Organizations and their Participation in Central Institutions

As we saw, the potential for change and development in each pattern of settlement was not governed solely by the features of the pattern itself or the social group which created and sustained it. It depended also to a considerable extent on the nature of the organization through which villages of a given type interacted, participated in central economic and political institutions, and mobilized support and resources. The differential growth and development of this organization is summed up in three areas: the characteristics of the different movement frameworks and their development; the organizational needs, orientations, and drives of the founders, which gave the impetus for the establishment of the movement frameworks and influenced their characteristics; and the potential of the organizations to consolidate and grow, and their functioning and success.

The Structure and Development of the Settler Organizations

The first point of comparison is the speed and the scope of organizational crystallization. Thus, both Hakibbutz Hameuchad and Tnuat Hamoshavim (moshav movement) were established *but a few years after the founding of their prototypes.* Ein Harod was formed as a prototype, after leaving the Labor Battalion in 1923, and Hakibbutz Hameuchad superstructure was established in 1927. Nahalal was settled in 1922 and the first committee of all moshavim met in 1926 (although the actual center was formed in 1930). In other words, within five or six years from the formation of the prototypes, countrywide movement frameworks were established.

Moreover, even before 1926–27, when the movements organized, the settlers of the labor sector (including movements not discussed in this study) did not lack the urge towards national organization. Thus, the small population of rural workers had been able to establish much earlier a national political and occupational body, the Agricultural Workers' Association; and this group was closely associated with the formation of countrywide political parties.

Later, the movements themselves, while still lacking an independent internal structure, had ties with one or another of the workers' parties. These, in their turn, tried to extend both moral and material support to the settlers. Hakibbutz Hameuchad had its close ties to a particular political party, Achdut Haavoda; and a similar drive (if less intense) was found in the moshavim and their links with Hapoel Hatzair party.[9] Parallel to this, the General Federation of Labor, established in 1920 and strengthened by the arrival of a wave of pioneering immigrants during the twenties, also began to play a central role.

The colonies, on the other hand, established their frameworks only some twenty-five years after the first moshava had been founded and did not create, even later, anything transcending local-regional or sectoral interests. Thus, the Federation of Judaean Colonies, founded in 1913, was not only a typical locally based organization, but also had a federal structure in which the component units were the colonies and the authority of the federation was, in fact, minimal. The Farmers' Federation, founded in 1923, more closely approached the organizational style of Hakibbutz Hameuchad and the moshav movement. The former federative basis of membership was abandoned and now became individual; the farmers were organized into agricultural committees. However, this federation did not include farmers in all the colonies: the farmers of Galilee did not join and never attained organizational scope comparable even to that of Judaea in the period prior to the First World War. Indeed, they continued to operate under the guardianship of the J.C.A. until the early 1940's, and only then began an independent existence. Furthermore, the organization of the colonies was not only regional instead of national in scope, but their attachment to national parties was also vague and divided. This was due to the lack of crystallization of the private sector of the country.

The difference in the developmental path of these three types also found expression in yet another area. Both Hakibbutz Hameuchad and the moshav movement quickly established complete hierarchical institutional structures, ranging from the conference to the small secretariats. In the moshavim these bodies operated more slowly and sporadically, especially since their functions were not clearly defined. However, by 1935 a number of decisions placed both the individual settler and the moshav committees under the authority of the movement center. This was done by the adoption of collective agreements designed to convert the local institutions into tools of the central body in all matters regarding farm planning and landholdings. In practice, the gap between the taking of such decisions and their implementation was, as we saw, great, and in this case, attempts to carry them out were not successful. A further difficulty lay in the fact that moshav membership was not homogeneous; during the 1930's many members joined because of instrumental considerations, and they were not prepared to accept domination in the ideological field and the consequent control of farming matters. For quite some time, indeed, the leadership was engaged in dealing with the crises that occurred in newer moshavim and in absorbing the heterogeneous elements into the movement framework. All in all, however, the principle and the practice of a central hierarchy became institutionalized.

The leaders of Hakibbutz Hameuchad were able to devote more attention to the development of inter-kibbutz economic cooperation because of the relatively greater degree of internal integration. This also facilitated the mobilization of manpower, both local and from overseas, and the opening up of new work opportunities for their members. Participation in the work of the Labor Federation was also intensified, as was the activity in the field of social and ideological communication. Such a wide range of activities reinforced the urge towards social and ideological autarchy, so Hakibbutz Hameuchad initiated an independent political party. Legitimation for separatism and splitting away was found in the conflicts within Mapai around foreign and home policies, which contradicted some of the basic values of Hakibbutz Hameuchad. The moshav movement also eventually adopted several of Hakibbutz Hameuchad's activities, with the exception of an independent political identity. However, while this delay and a different scale of priority for different levels of activity was to determine the different developmental path of the movement as a whole, both workers' frameworks contrasted strongly with that of the farmers' organization, which throughout its existence had been concerned primarily with purely economic matters, and neither regulated and mobilized the colonies internally, nor integrated them into the general political system. In fact, it began to appear as a separate political entity only in the thirties, when its activities alternated between, or combined, marginal efforts within the Yishuv institutions and the external mandatory authorities.

In conclusion, all three settlement organizations were characterized, at least in their advanced stages of development, by an increase in numbers, internal institutional differentiation, and a widening of their range of operations. However, these similarities were external, and it is impossible to ignore differences in paths and rates of organization. Thus, for example, the tempo of centralized organizational development and range of operations was extremely slow for the colonies, and in fact, they did not overcome their initial delay until the close of the period under review. The moshavim, which had developed an organizational framework at an early stage, failed to give it adequate support and so lagged behind the development of Hakibbutz Hameuchad. Moreover, these differences function not only with regard to tempo and range of development. Variations exist also in the degree of control exercised by the movement center over the periphery, with the colonies at one pole with the most limited range of authority, and Hakibbutz Hameuchad at the opposite pole, with the moshav movement located between these two, but approaching the pattern of the former rather than that of the latter.

Hakibbutz Hameuchad also has the widest range of activities in all institutional spheres. The colonies, in all their organizational set-ups, almost always emphasized the economic interests of the monocultural farms, and while political activities were sometimes undertaken, they were usually short-range, unplanned ones. The moshav movement, whether of their own accord or of necessity, gave up, at the beginning, certain of the functions adopted by Hakibbutz Hameuchad, such as the occupation of economic power positions in the private or governmental sectors of the economy or the crystallization of a separate political outlook. With regard to other functions, however, the movement stood closer to Hakibbutz Hameuchad than to the colonies. This attitude was essential because of the struggle against organizational and ideological deviations.

We must now turn to the factors which gave rise to this differential development during the process of organizational crystallization; and first to the nature of the initial drives and objectives which led to the establishment of the organizations.

Organizational Drives and Objectives

As we saw, the farmers' organization appeared relatively late, had a less articulated structure, represented narrower interests, and branched off into fewer activities. Secondary, though significant, differences existed also between the two workers' movements themselves. These variations could not, as was seen, be explained merely by the different political circumstances surrounding the founding of the three settlement types. True, the colonies were initially under a system of administrative guardianship that limited and delayed the need for and the possibility of a supra-local structure. Furthermore, the farmers' organization made its first steps under Turkish rule, when political activity was very restricted. However, it was under the same political conditions that the small population of workers, also rural rather than urban in character, was able to develop a much greater initiative. Nor was the objective economic incentive to organization greater in this case, since the resources available to the newly created Zionist institutions were then very poor.

Time sequence is still less capable of explaining differences within the labor sector; and while the movements, once established, generated and sustained their further development, the initial factor was that of differential organizational drives, orientations, and potential. These determined to a large extent the organizers' tactics for uniting people with similar interests and also shaped their demands. Furthermore, the quality of such conceptions influenced not only the nature of demands, and of the expected rewards, but also the degree of readiness and ambition to shape the nature of central institutions that had the monopoly over political and material resources. The present section compares the characteristics of the different founding groups in this respect.

The farmers of the colonies, when they established Pardess and Agudat Hacormim, restricted their demands to a limited institutional sphere and aimed at economic freedom of action after a period of bureaucratic guardianship. With the foundation of the Federation of Judaean Colonies, a number of years after Pardess and Agudat Hacormim, the operational objectives were widened to include the defense of the few political rights enjoyed by the Jewish settlers under Turkish rule. However, this was manifested in terms of personal influence exerted on various regional governors, and was of necessity characterized by bargaining between the leaders of the farmers and the holders of bureaucratic authority for privileges and favors. This pattern was thus similar to that found in many colonial countries prior to the establishment of a national movement.

The motives of the Farmers' Federation, founded in 1923, were somewhat different in character. Certain of the planks of the federation's platform explicitly mention the need for organizing the farmers of the colonies as a political body, so as to assure their appropriate representation in the relevant institutions, especially vis-à-vis the General Federation of Labor, which had been founded two years earlier. However, above the desire to establish a political body of farmers, the economic needs were still the most important; the main aim was to turn the federation into a tool for the development of the farms by encouraging cooperation in supplies, marketing, and credit. Moreover, in spite of the appearance of an explicit political dimension as a partial motivation in establishing the superstructure, the concept of political representation and participation here had a different connotation to that held by the leadership of Hakibbutz Hameuchad and the moshav movement. Although the federation, founded in 1923, was much more universalistic than its predecessor, founded ten years earlier, it still had a sectoral and localistic character. This was expressed in many ways. For instance, the Farmers' Federation strongly opposed the policies of the countrywide political institutions, such as the Jewish Agency Executive or the underground military organization of the Jews (the Hagana), whenever these ignored economic or local interests. The farmers, dependent on cheap Arab labor, were justifiably sensitive to the lack of consideration shown by these institutions at times to the delicate balance in Jewish-Arab relations in areas where the colonies were located.

A further example is the demand made by the federation for a separate parliamentary representation in the Assembly of Representatives that was to take the form of a curia. The grounds for this demand were, it will be recalled, that the economic share of the colonies in the national product was greater than that of other sectors. The rejection of this demand led to their boycotting of the elections, which further weakened their position in the Jewish political structure.

All in all, the farmers were largely marginal to the ideology and the aspirations of the national movement and institutions, and their political demands were consequently directed to gaining influence with the British administration, which the Jewish sector of course considered a center of external and hostile power. The main effort was thus to obtain taxation and customs relief from the mandatory authority and to give a legal basis to the agricultural councils, so as to limit the active and passive voting rights to permanent settlers and to restrict the political power of the workers. Because of this, they also attempted to obtain control over a part of the immigration quotas and in this way to regulate recruitment of manpower in the colonies and to reduce to the minimum their dependence on the General Federation of Labor.

It is undoubtedly true that, in certain respects, both Hakibbutz Hameuchad and the moshav movement were also sectoral in character. This was especially felt in the struggle with the farmers of the colonies and with the political adversaries of the Labor Federation in the cities. Essentially, however, the pioneering settlements, and in particular the kibbutzim, conceived of their movement in terms of national needs and interests, and saw their way as the best—if not the only—means of developing and regenerating the country and the community. The very act of organizing arose, in this case, from a sense of responsibility by a social-political advance guard and a potential center of a mass movement, which rejected the idea of any contradiction between its own specific interests and the national ones. This claim, which was indeed accepted by most of the Yishuv and the Zionist movement as a whole, determined the tactics and the strategy adopted by Hakibbutz Hameuchad, and to a lesser degree by the moshav movement, in the attaining of their objectives. It led to the occupation of key points and bridgeheads in the great human reservoirs of East European youth, and to the penetration into general economic activities of the country in colonies, industrial plants, ports, and construction and road-building.

It is important to remember, however, that in these objectives it was Hakibbutz Hameuchad élite which led the way, while the initial motives of the moshav organization were more specific and practical in character. Indeed, the settlement institutions of the Zionist movement preferred to deal with groups rather than individuals, and the moshav members had little choice but to follow suit. From this point of view there is perhaps an actual similarity between the foundation of the moshav organization and the sectoral organization of the colonies. The moshav movement was formed mainly in order to compete for financial resources, land, and manpower, and to fight for equal standing with the kibbutz; its actual expansion and spread came only later and was more the result of spontaneous growth than of any organizational effort.

Controlling and consolidating this mass movement formed one of the central functions of the moshavim frameworks; the necessity to combat deviations from the pattern of Nahalal commanded most of the effort. In this respect, however, the moshav movement was different from both the colonies and the kibbutzim. In the former, because of their permissive nature, struggle against ideological deviance was irrelevant (except, early, in the sphere of religion); while in the latter, the problems which arose during the late twenties and thirties, and even the forties, were much more marginal and constituted less of a social problem. This indeed serves to explain the efforts made by the moshav movement to establish frameworks for the taxation of settlers and for the registration of the villages as cooperative societies responsible for public property and internal working arrangements, which would give the organization the appropriate legal tools. As will be recalled, the movement was only partially successful in realizing this plan, which has recently reemerged in the form of the Moshav Bill. Another weakness, especially in the light of the rapid growth in numbers, was the absence of planned manpower recruitment. This was in part due to organizational problems, but it also resulted from a failure to overcome the contradiction between the desire to attract additional members to moshav settlement and the lack of capacity or willingness to train such manpower in veteran settlements, mainly because of the principle of self-labor.

In general, we can say that the organizational objectives of the Farmers' Federation were limited, at least initially, to economic aims. From time to time other objectives did arise, especially in the sphere of national politics, but even in such cases the demands were narrow, specific, and particularistic, and directed not to the elected bodies of the Jewish community but rather to the external British administration.

Hakibbutz Hameuchad contrasts very sharply with this position. From the beginning, its organizational objectives included with equal intensity and power all institutional areas. The demands made were mostly universalistic and diffuse in nature, and while they showed traces of a sectoral emphasis, there was nothing parochial about them.

The moshav movement again fell between these two poles. Its organizational orientation was national and potentially included a broad spectrum of interests in such areas as mobilization of resources, ideological indoctrination, and education; but in practice crystallization of these aims was slow and accompanied by many doubts and difficulties.

These, then, were the ideological predispositions, as it were, which gave the various settlement organizations their initial impetus and influenced the characteristics traced in the preceding section. Attention must now be focused on differences in organizational processes, channels, and struggles.

Organizational Activities, Channels, and Achievement

As was seen above, the social and ideological characteristics of the different founding groups were a fundamental factor in the differential spread and growth of their settlement patterns. The pioneers of the kibbutz thus conferred upon the future movement the potential advantages of an expansionist drive and orientation, as well as a social structure and force geared to the particular challenges and problems that would face the country, in particular those resulting from the security situation. Predispositions, however, were no more than points for organizational take-off; and no less important have been the mechanisms which enabled these qualities to be realized and modified, and which facilitated sustained and unflagging organizational growth and innovation. The functional "fit" of the kibbutz to national requirements and conditions was neither easily accepted nor in fact inevitable. Indeed, the collective fought a hard up-hill battle for recognition, both within and without the workers' movement. Nor was it for a long time the clear choice from the point of view of numerical strength, an area in which both Hakibbutz Hameuchad and the moshav movement lagged behind the colonies. Indeed, up to the early 1930's, the population of the two movements together was only 1,500, as compared with 5,000 landholding farmers in the colonies, and only by the mid-thirties did the situation become reversed (in 1936 there were thus some 7,000 *farmers* linked with the Farmers' Federation, although the total population of the colonies, including hired laborers, was 56,000), while Hakibbutz Hameuchad had 6,500 adult members and the moshav movement 3,000.[10]

Moreover, the priority given to the kibbutz in the early thirties because of economic and defence considerations did not mean that the collective was the only alternative as a *pattern*. It is quite feasible that, given a Zionist Executive of another political complexion or conviction, emphasis would have been laid on the moshav or on private farming, to which funds and manpower could appropriately have been directed. While such a choice might not have been logical or efficient in objective terms, it might have been pushed through and upheld on a political basis. The dominance of the kibbutz was by no means historically pre-determined, as it were, because of its potential advantages. It was the growth of the power of Hakibbutz Hameuchad (and of other collective movements) that affected, to no small degree, the actual pattern of rural development, which might have taken another path. In other words, Hakibbutz Hameuchad had to be able to realize the collective promise by ensuring at least three things:

Firstly, that the potential respondents to demands and needs should be aware of the existence of such demands.

Secondly, that most of the demands should be translated into actual policies and resources.

Thirdly, that the organization should assure the capacity to absorb and utilize the resources made available.

A comparison with the other organizations will show how it did so. As we saw, up to the First World War the farmers of the colonies operated almost without competition in the internal political and economic spheres. By dint of considerable flexibility, adaptability, and initiative they overcame manifold difficulties, gained economic power positions as well as political "connections," and achieved social status and esteem. This was especially so after they escaped the guardianship of Baron Rothschild. However, by the second decade of the twentieth century, the colony farmers had begun to lose much of the prestige they had enjoyed among Jews in Palestine and abroad. They were accused, not always with justice, of being materialistic, indifferent to the national movement, and unwilling or unable to respond to the new challenges of land settlement and agriculture.

By the end of the First World War, the farmers, while retaining considerable economic power, had lost much of their political influence. Not only their aims and ideology but their bargaining techniques became an anachronism in the new conditions created by the Balfour Declaration and other political achievements of the Zionist movement, as did the regional organization typical of the farmers, which greatly restricted the possibilities of manipulating manpower and financial resources. It was indeed only later that the farmers drew the appropriate conclusions and established the Farmers' Federation on a countrywide basis to counter the General Federation of Labor. The crucial difference between the two bodies was not only that the one was the innovator and the aggressor, while the other only reacted and defended itself; no less significant was the fact that the Histadruth combined and commanded the resources of a whole rural-urban sector, while there was no such crystallization on the right. Moreover, the moshavot were marginal also in the capitalist sector itself. The settlements of the workers' sector, on the other hand—and in particular the collectives—consolidated their central position in it, and mobilized its resources and support even in areas in which opinions differed. This success was neither easy nor obvious, and the struggle of Hakibbutz Hameuchad to obtain key positions in the Federation of Labor and its constituent parties was indeed the most vital factor in subsequent development. Some examples recollected from the material presented will doubtless bring this point forcibly home.

Development of frameworks for recruitment and training of manpower both in Europe and in the country. The kibbutz thus had a dominant position in attracting pioneers from Hechalutz movement and a virtual monopoly (together with the Hashomer Hatzair) in local youth movements.

Widespread cultural and ideological activities, such as the organization of regular seminars and courses and publication of books and periodicals. The intensity of these activities is the more striking as the population concerned numbered, by the close of the period of this study, no more than 10,000 to 12,000 adults. Only supreme confidence in future growth, together with a sense of mission, of uniqueness, and of political and moral superiority, enabled these settlers to carry the heavy economic burden of this propaganda. The moshavim lagged behind the kibbutzim in this respect, while in the colonies such activities were almost entirely absent (although purely occupational communication was widely developed).

Economic and political entrepreneurship and public image. In this area, the organization took great pains to help the collective settlements to overcome difficulties and growing pains, and to drive home, to stubbornly disbelieving audiences, the proof of their ability to survive and to expand, and to do so under hard conditions and scarcity of resources. This agrarian activity and salesmanship was accompanied by demonstrably successful mobilization of manpower for general pioneering tasks, such as the building of the Naharayim Dam for the electrification of the country, the Dead Sea Potash Works, or the Haifa port. The importance of these achievements lay not only in tackling the jobs, but even more in the penetration into all levels of work, including the lowest, usually reserved for unskilled Arab labor. In this area the kibbutz spearheaded the workers' movement as a whole.

The moshavim were no less successful than kibbutzim in the area of internal economic affairs, but they were far behind in this sphere. Here the kibbutz reigned supreme, the moshavim being faced by several obstacles, over and above their less militant ideology as such: (a) the lack of wide consensus and solidarity in the movement, due to the increase of social heterogeneity; (b) the agrarian system based on the family unit, which made mobilization of manpower difficult for internal, let alone external, public activities; and (c) a passive role played in the Histadruth, in Mapai, and in the Agricultural Center, which contrasted sharply with the refusal of the kibbutz to accept decisions handed down from above, and with its unceasing attempts to gain dominance in all accessible centers of power.

Indeed, since the supreme Zionist institutions could not then be directly influenced, the kibbutzim concentrated on penetrating the Histadruth and the party—first Achdut Haavoda and later Mapai—whose representatives had access to the highest echelons and eventually gained decisive influence in them. This influence was never based on numerical strength, and none of the settlement movements ever commanded a majority in any of the party or Histadruth organs. It reflected rather an advantageous bargaining position based on the often emphasized moral superiority of being closest to realizing the national ideal and of representing best the interests and ideals of the workers' movement. This superiority became, of course, most pronounced—morally as well as politically—with the growing initiative and the central position of Hakibbutz Hameuchad in national defense, and its almost complete monopoly in the provision of manpower, training, and bases for the Yishuv élite corps. All this contrasted most sharply with the farmers of the colonies, who failed entirely to gain such an evaluation from either the élite of the workers' movement or from large sections of the private sectors of the economy.

No less important was the fact that the kibbutz movement as a whole, and Hakibbutz Hameuchad in particular (and to a lesser degree the moshavim), succeeded in establishing an organizational framework for the allocation of material and manpower resources. In this the settlers generally *preceded* the settlement authorities and forced them to adapt themselves to the already existing frameworks. These eventually became extremely efficient in channelling resources and in determining the settlement authorities' policies. Such a framework is to be found in the complex structure of Chevrat Haovdim of the Histadruth, with its special financial instruments such as the Workers' Bank, Nir, Tnuva, Hamashbir, and such political organs as the Agricultural Center and Hechalutz youth movement; and to these must be added the organizational set-ups of Hakibbutz Hameuchad and the moshav movement themselves.

The farmers in the colonies proved themselves to be helpless in these respects, even in the period prior to the First World War, when they had a monopoly in the area of normative ideals. That is why they were forced to accept the system of receiving aid from above, even when this meant the loss of internal autonomy and damaged the farmers' morale and their economic conditions. It was only after economic initiative was transferred from the external bureaucratic élite to an internal élite that the farmers were able to recover morale and improve economic matters. But since this recovery was not directly due to the activities of the settlement institutions of the Zionist movement, the farmers' failure to influence the nature of the Zionist movement did not hamper them in economic affairs. The dominant economic branch of the colonies since the close of the First World War was citrus fruit, mainly destined for export. The colonies were thus less dependent on internal markets than the kibbutzim and moshavim. They were also less affected by the dynamic development of the Jewish sector that occurred with increased immigration after 1919. For them, it was essential to maintain close contacts with the mandatory government, since the latter had exclusive control over imports and exports. However, the failure to develop communicative channels with the internal political power centers proved fatal when the farmers attempted to translate their economic power into equivalent positions of internal political strength. They were not only denied political weight, but their undoubted pioneering achievements and birthright, and their contribution to development, absorption and employment, were negated and overlooked.

The study summarized in this chapter presented three stories of rural development, which together give an idea of the checkered cloth from which Jewish settlement in Palestine was fashioned. The interesting fact was that, though their temporal sequences of development were different and though they represent three distinct patterns of rural life and of organization, they all produced modern farming and village systems. It is, of course, impossible to evaluate the relative merits and advantages of each such system; and our purpose has been rather to highlight their special characteristics in terms of fulfilling their different aims, of solving their tensions and institutional problems, and of developing in a sustained way. As we saw, the three configurations do show significant contrasts in these respects. None of them, unless one considers the process of urbanization a mortal sin, evince any crippling disability; by and large, the three villages are all a study in success. To put it more precisely, they all give scope to their particular aims and qualities, without paying the price of social or economic failure, and without becoming a liability. A much less balanced view is, of course, seen in the sphere of voluntary settlers' organizations, where the initiative, scope, and intensity shown by the collective movement are outstanding. This, and other issues of the study, are the subject of a few concluding remarks, in which some general insights obtained on nation-building and rural development are suggested.

9

General Conclusions

In the preceding pages, we have examined the major trends of rural colonization in the Jewish community of Palestine, adding some remarks on the situation after independence. The scope of the discussion was strictly limited, focusing on selected types of settlement and their differential characteristics and problems, and it did not analyze systematically the development of the rural sector as a whole, or its relative place in over-all social processes.

Development in Palestine took place under enormously difficult, truly pioneering conditions. Hardships resulted from sickness, in particular from malaria-infected swamps and the climate, to which the newcomers from Europe were unaccustomed. Security conditions were consistently bad, with raids on life and property; this not only affected the individual, but also constantly drained resources and imposed constricting defense considerations. To top it all, Palestine was a country poor in natural resources in general and in agricultural ones in particular, so in many areas, farming had to be undertaken on insufficient and low-quality land, and without enough water.

On the positive side certain unique circumstances favored colonization of the country. First, the framework for development and mobilization of resources extended far beyond the territorial limits of the society. It was thus able to continuously tap outside sources for enlisting support, for accumulation of capital,[1] and—most important perhaps—for recruitment of highly educated and skilled manpower. This mobilization was, moreover, driven by the immense ideological upheaval created by the historical circumstances of the Diaspora, which stimulated enterprise, innovation, and pioneering. These ideals were translated into practice relatively easily because the process of development started in many spheres with a clear slate.[2]

Nor was political dependence a wholly negative factor, from the point of view of nation-building, at least from the twenties onwards.[3] Here was a colonial people who achieved a large measure of autonomy and political organization very early, and who could therefore legitimately build their institutions and strength from the ground up, free from sovereign responsibility over such spheres as foreign relations, the monetary system, infrastructure, and basic services. The fact that the Yishuv operated under a dual political system and had only a partial institutional structure of its own created many restrictions and difficulties in immigration, in purchase and colonization of land, and in security, which the government was not always willing or able to provide. On the whole, however, the balance for national development was positive,[4] and the various challenges kept the community on its toes and in a state of constant mobilization and creative tension.

In spite of its narrow scope and highly local color, the material presented has a more than parochial significance. Some broader insights into processes of development and modernization suggest themselves at three levels: in the rural community, in the agricultural sector, and in the nation and society as a whole.

Some valuable insights appear, first of all, regarding development of the farm and the farming community. In our introduction we referred to the problem of the range of structural and attitudinal variability consonant with development; but the issue is more often spelled out in negative terms, namely, the extent to which different social factors impede development. Hirschman faces the question most incisively, although he deals with it more on the macro-level; he declares himself squarely against the obstacle-hunting approach in general, maintaining that he has yet to come up against an apparent social barrier that can neither be gotten around in the process of change or is not later found to have been an advantage after all.[5] In some important respects our data, though limited, support this rejection of the often unfruitful preoccupation with difficulties; they show that there are many different paths to development, and in each pattern a range of functional equivalents allows it to "neutralize" or bypass "inconvenient" spots. Thus, in the moshav, the most formalized and inflexible of the forms analyzed, producers' subcooperatives could function as alternatives to nonlegitimate large production units, allowing the small-holding farmer both to specialize and to enjoy the actual advantages of size. At the same time, Hirschman creates a dangerous illusion in two important respects. First, while it is no doubt true that most *single, specific* obstacles can be compensated for in any system, a combination of several factors acting together may prove too much for it. Second, he takes for granted the presence of a drive to manipulate or change institutional arrangements in order to find solutions to problems of development—a factor that is by no means self-evident, and whose absence is in itself perhaps the most crucial obstacle.

As our data clearly indicate, each pattern, however functional and advantageous at any given time, sooner or later faces the necessity of change. Development is not—as Rostow seems to have hinted [6]—self-maintaining once take-off is achieved; a sustained process clearly requires constant re-evaluation and innovation, without which stagnation or breakdown of development are bound to occur. The case studies presented thus support more general findings on the societal level.[7] They also show, however, that the social mechanisms of experimentation and change fall into two essentially distinct categories. On the one hand, there is a search for new solutions and new achievements *within* the limits of the fundamental provisions of the system. Both the collective and the cooperative village belong to this category; and it is in these terms that the apparent difficulties now facing the latter form can essentially be spelled out. The moshav, unlike the kibbutz, seems to have largely exhausted the legitimate range of secondary institutionalizations, *relatively limited in its carefully worked-out pattern*, and is unwilling to take steps beyond the essential commitments. In the colony, by contrast, change *beyond* institutionalized arrangements and commitments is legitimated in the pattern itself, thus representing, much more than the other types, not just change but *modernization* in the specific sense.

This difference between controlled change and the free play of motives and solutions is indeed the heart of the matter presented, and the paradox of modern rural life. For while change in the sense of modernization is initially a prerequisite of development (and both the collective and the cooperative had originally been completely new experiments), above a certain level, rural life cannot give it free scope if it is to maintain itself. The great achievement of the kibbutz is that it has been able to develop *and* to preserve its identity. As we have seen, however, this form is socially expensive and unlikely to provide a prototype. The greatest challenge of rural development is to evolve other patterns for developing while controlling modernization.

So much, then, for the developing community. The colonization of Palestine also throws light on some basic issues of rural development as a sector. A major suggestion, *though only a cautionary one*, emerges regarding the choice between development based on steady, step-by-step growth, as against a process of dramatic spurts, in which advances are made into new areas and enterprises before earlier projects are consolidated. As we saw, the moshav and the kibbutz represented the two conceptions respectively, and the over-all decision was taken in favor of the latter's dynamic approach. Under the circumstances, this seems to have been the proper course. Our data, though, suggest that one should be wary of the risks of failure involved in the "great leap forward" vision, *unless the stability of the over-all institutional structure and the social security of the new communities are well established.* In the situation described here, both conditions existed, the second especially being assured and promoted by the organizations of settlement movements, as well as by the great cohesiveness and ideological commitment of the settler groups. With this factor decisively weakened, however, the vulnerability of villages exposed to long periods of half development and partial consolidation appears to be dramatically increased. Thus, in Israel itself, in spite of the greater resources and the backing of the State, the new immigrant moshavs faltered under the same policy.[8]

The range of the developing rural sector is another area on which the material seems to throw a general light. In this respect there are indications of the favorable impact on development of agrarian pluralism as such, with a variety of patterns coexisting and competing with each other. Of course, the presence of different systems, connected with the market on the basis of different assumptions, is likely to create economic strains, as is also the struggle for scarce resources. By and large, however, diversity prevents all eggs from being placed in any one development basket, each of which, as we have seen earlier, is likely to wear thin in time.[9] Diversity promotes social experimentation, innovation, enterprise, and rural mobility, and eliminates a monopoly of one form, one ideology, and one interest aggregation. This consideration brings us directly to our last topic, the implications of the case studies on the level of the developing society as a whole.

The growth of Jewish agriculture in Palestine is a decisive factor in the transformation of the historical Jewish society. It is true that in our case we cannot speak of "nation building" in the crucial sense of creating a national consciousness; from the beginning, collective identity was strong and unswerving. Nevertheless, the story of the rural settlement in Palestine doubtless constitutes a focal part in the development and the modernization of the society as a whole, in terms of intensive structural differentiation, of the abandonment of traditional roles, and of the assumption of new ones in economy, science, politics, the arts, and other spheres. Moreover, the events described above are generally relevant not only to the understanding of modernization in the sense of structural and demographic differentiation, but also as regards the creation of mechanisms for *sustained growth*, which enabled it to develop after the initial impetus or take-off and to initiate and absorb change and innovation continuously. The capacity of a society or subsociety for such sustained growth, for modernization in the essential sense of legitimating and providing for new goals and structures *beyond* institutional arrangements existing at any given time, has recently become a major focus of sociological theory and research on development.[10] And although the problem in general is clearly beyond the terms of reference of the present study, one aspect at least lends itself to an analysis: the relationship in development between the central political and ideological institutions and the rural sector, or, more generally, between *center and periphery*.[11]

The integration of the center and the periphery requires constant regulation in any society, but the need for it is particularly acute in a situation of far-reaching structural and demographic changes, and in a context of particularistic groups—territorial, ethnic, and other—with autarchic tendencies. Here the degree of identity and of simultaneity in the changes occurring in the center and periphery, and of the extent of their mutual adjustment over time, are of critical importance. In fact, the ability of a society to integrate and activate the peripheral groups and to mobilize and reallocate their resource potential, whether in the form of manpower, goods, or political support, is a crucial test of its capacity for sustained growth. Such mobilization and reallocation of traditional and of dispersed resources—essential features of the process of modernization—involves of course the dissolution of privilege and the break-up of monopolies; it can be defined, by broadening Polanyi's classical economic term, as the *unfreezing of resources*.[12] In one place the frozen resources may consist of land that cannot be reparcelled or transferred; in another it may be manpower prevented (subjectively or institutionally) from sectoral or outside mobility, vertical and horizontal; while in others still political or symbolic overcommitment impedes changes in the bases of social consensus and the legitimation of new élites. Different historical societies and phases of development can, obviously, be characterized by different profiles of "resource-freezing"; but the ultimate test of the capacity for sustained growth of a society, and of the viability of its political center, is surely the ability to meet these problems and to create a *flow of "free-floating resources,"* [13] economic, political, and symbolic.

The relevance of the case studies presented lies in their demonstration of the crucial role played in this respect by the rural sector, which proved a vital factor in the modernization of the society in general, both through the supply of economic resources and through participation in the creation of a viable political center. As could indirectly be gleaned from the material presented, the central institutions of the Jewish community in Palestine very early possessed many crucial characteristics that distinguish a modern political system capable of meeting the problems of sustained growth.[14] There occurred, for example, internal differentiation within the political system, hand in hand with a process of unification and centralization. This process included a rationalization of authority, the emergence of a legal-constitutional framework, and the creation of a more or less efficient public administration; at the same time, there was an ever-growing mass participation,[15] associated with the growth and articulation of "free" political power. No less significant was the promotion of leadership with the ability to internalize the rules of the game so as not to abuse power and privilege, and with an initiative to penetrate and mobilize the newly emerging political crystallizations.

As was mentioned earlier, the Jewish community in Palestine differed from other developing societies, for it was a transplanted one, a society institutionally and ideologically much less frozen, in which the successive waves of newcomers constituted themselves, as it were, free-floating resources, while the mandatory government provided the requisite infrastructure. The interesting fact to emerge, however, was that the representatives of the rural sectors were here among the most important political entrepreneurs and were particularly active in the building up of institutional frameworks and in the articulation and the mobilization of various interest groups, calculated to assure the optimal flow of resources and norms between center and periphery.

Thus, we see that the Jewish rural settlement in Palestine was characterized by intensive, even feverish creativity, in the course of which an almost completely new society and a modern agrarian sector developed in the span of one generation. This sustained effort depended largely upon the continuous mobilization from outside of considerable resources in funds, skilled manpower, and ideological commitment. Over and above these, however, the process of colonization and of nation-building in general was here shaped and facilitated by *a special kind of relations between the rural sector, or periphery, and the national center—relations which embodied two main features:* simultaneity of development, i.e. the growth of rural society was concurrent with the crystallization of modern central institutions; and a high measure of close interaction and support between the two.

This comprehensive pattern of nation-building rested upon the fact that neither the center nor the periphery evolved organically, or directly from traditional structures. Moreover, both were fashioned by the same national revival movement and thus constituted but two manifestations or aspects of the same drive. Hence also the multiplicity of ties and of common aims: a sharing of values, a partnership within the same economic, organizational, and political frameworks, a sense of social integration and belonging, and even a large measure of identity of central and rural élites. The interaction thus created in each sphere was most intense; and this intensity was further reinforced by the scope of relations, as well as institutionalized by an organizational framework binding together village, movement, party, and center. The direct access of the rural sector to the supreme national goals and institutions, based on it being one of the main upholders of these aims, together with its close ties with national leadership, gave it a position of autonomy and power on the one hand, and a sense of responsibility and partnership on the other. In consequence of all these, there was, at least from the thirties on, a constant two-way flow of resources—manpower, capital, produce and goods, political support, social solidarity and legitimation—between the rural sector and the center, with the intensity of relations sustaining the flow, the comprehensive organizational framework regulating it, and the position of the rural sector safeguarding its reciprocity.

It was this symmetry of ties and responsibilities that enabled the rapid economic growth of modern agriculture and agricultural industries and which made available to the center much of its requirements, especially in manpower and legitimation. A similar give and take took place on the symbolic level: the élite of the settlements was thus a source of regeneration and creation of common values and a new way of life, while the center made available to it the tools and channels of activity in the society as a whole, with both contributing in this way to the crystallization of national identity, the establishment of educational patterns and institutions, and others.

It thus seems that the aim of intensive growth and modernization was quite well served by a pattern of development characterized by close and reciprocal ties, encompassing both *resources* as well as *initiative*.

The special pattern of development, created in Palestine and based on the simultaneous growth of the reciprocal ties between the national center and the rural sector, proved a powerful instrument for building up both. This reciprocity must itself be capable of changing in time and of keeping open and flexible; as we have seen, there are potential obstacles to such a reshaping inherent in the pattern. Some of these obstacles have indeed been further emphasized and intensified by the establishment of the State.[16] For example, the fact that the system has been geared to intensive mobilization of resources and to close interaction between the rural sector and the center, thus promotes the process of subordination of hitherto independent units to large corporate bodies, or the creation of corporate hierarchies descending through all levels of activity. In this way the Histadruth and the political parties control movements, movements control communities, and the individual settler is controlled by all—with the hierarchy constituting the main channel of allocation of resources. Thus, while organized mobilization for development worked its wonders in Palestine, and sparked an unprecedented surge, it may also inhibit to some extent individual and community initiative and internal innovation.

Another corollary of the organizational set-up described may of course be a measure of monopolization of agrarian forms. As we mentioned earlier, rural diversification—spontaneous experimentation and creation of new forms—had been one of the great advantages of the settlement process analyzed. Crystallization of powerful vested interests, however, may later impede new initiative and a free flow of resources, hamper the emergence of additional novel forms, as well as inhibit the development of some of the existing ones. The intensity of ideological and political commitment constitutes potentially still another built-in disadvantage. Such commitment was, of course, the essence of the immense upheaval in the wake of which development prospered. As we saw, however, it also gave rise to the formation of uncompromising and unyielding attitudes and interests, which sometimes prevented cooperation and orderly relations between groups even in the instrumental spheres and contributed to splits and cleavages on party, movement, and community level.

Intense identification with and participation in central national aims and institutions may affect not only the rural sector, but also relations with the center, and the center itself. This seems so because the initial primacy of this sector, and its having been largely representative of general national and social aims, may create in it a *blurring between identification with such general aims, and between insistence on identity of interests.* Initially, the two were largely one, and the good of the settlements was also, ultimately, the general good. In time, though, there inevitably arise competing aims, sectors, and élites, and the rural sector may find it difficult to recognize their claims and to adjust and subordinate itself to them. By the same token, the center may be over-identified with it, at the expense of industrial and urban groups and interests.

By and large, however, these obstacles are being overcome, and on balance, the mobilization potential of the pattern described is also today a valuable force, as exemplified by the establishment of the new immigrants' moshav.

However, this certainly does not mean that the pattern of development established in Palestine *as such* would be suitable or even workable in other social settings. Besides the many other differences between events in Palestine in the first half of the twentieth century and the conditions in the developing countries at present, one decisive difference is the nature of the mobile social forces with which the political élite has to deal. In Palestine, the keynote was *associational interest groups*, even if possessed of a strong ideological and missionary tinge; while the crystallization of rural groups in the developing countries today, in so far as it achieves any articulation and integration, bears the stamp of *nonassociational interest groups*, groups built on ascriptive and particularistic criteria, whether lineage, tribe, caste, language, territory, or others.

Moreover, in not a few cases, these organizations take the form of *anomic interest groups*,[17] especially when bitterness accumulates while legitimate—formal or informal—frameworks for protest are either non-existent or too rigid to absorb the demands. It is thus due to no mere chance, or to national idiosyncrasies, that the Palestinian pattern of relationship between village and city, or between periphery and center, seems to be quite rare in the history of agrarian and general modernization, whether in Europe or in the developing countries of today. Especially rare is the phenomenon—stressed repeatedly in this work—that it was not only the center which tried to change the rural sector. On the contrary, demands for change, reform, and innovation came often and even regularly from the rural sector itself. More frequent has been the opposite pattern of development, that of regulation and resources radiating from the center, examples of which are Japan in the end of the nineteenth century and the beginning of the twentieth, Soviet Russia in the twenties and thirties, and China more recently.

What usually characterizes developing traditional societies, however, is that the agrarian sector supplies, almost unilaterally, services and commodities (in the form of produce, rent, and taxes) to the urban sector and the political center.[18] Moreover, in some countries the major part of investment designed for development in general comes out of the capital accumulation of the rural sector. A good example of this structure is of course historical Japan, where agriculture was the main source of financing the Meiji regime.[19] The financial burden imposed on the Japanese farmers (which, by the way, was accompanied by an astonishing rise in productivity) was economically almost one-sided, while the center provided the initiative, administrative arrangements, and a uniform system of taxation.[20] As in the case of Palestine, the successful functioning of this pattern of development was made possible by, and in fact grew out of, particular social conditions and structure. The process of modernization could here be controlled and regulated by a small oligarchy, as it was not subject to serious conflicting pressures or strong political demands from below. In other words, this oligarchy functioned in a setting lacking revolutionary expectations, and could be quite successful in mobilizing traditional resources and in regulating and channeling potential free-floating ones.

By contrast, in most developing countries today, a double obstacle exists to the successful repetition of the Meiji relationship between center and periphery, that of an old-established élite establishing the goals and the infrastructure of development and using most of the resources, and that of the traditional rural sector providing much of the sinews. First, most of the political élites which grew after independence lack integration, unity, and that powerful authority emanating from the Japanese emperor and traditional national symbols—a factor that affects the feasibility of the Russian-Chinese pattern. Second, the quantitative growth and the qualitative expansion of demographic and economic pressures and political demands from various groups in the periphery is toward the center. In such a constellation, controlled and regulated change, based on partial and selective mobilization of material and manpower resources, seems doomed to failure from the outset.

To sum up, no historically successful patterns of development can *per se* be adopted as models for present-day (or future) societies, even though they may seem to have functioned better than many existing systems do. In the present work we have analyzed the relationship between development and social factors chiefly as regards agrarian forms, the organization of the rural sector, and, to a lesser extent, the relations between center and periphery; but there is no doubt that the truism holds also for other areas. The crucial question thus is which patterns of development "fit in" and can best be made to work under different social situations. It can only be hoped that patient and systematic comparative research will provide at least partial answers to this question.

Appendix: Chronology

1878 Foundation of Petach Tikva.

1881 Pogroms in South Russia; beginning of Hovevei Zion movement.

1882 Beginning of First Aliya.

1883 Baron Rothschild begins to support colonies; re-establishment of Petach Tikva.

1884 The Katowitz Conference of Hovevei Zion.

1887 First Organizations of Workers in Colonies (Rishon Le Zion).

1891 Ahad Haam's article: Truth from Eretz Israel; establishment of J.C.A.

1896 First Zionist Congress; establishment of the Zionist Organization.

1900 Transfer of responsibility for colonies from Baron Rothschild to J.C.A.

1901 Establishment of Jewish National Fund—Keren Kayemeth.

1902 Herzl's Altneuland.

1903 The Kishinev Pogroms in Russia; sixth Zionist Congress and the Uganda Dispute; opening of the Anglo-Palestine Bank.

1904 Beginning of the Second Aliya.

1905 Abortive revolution and pogroms in Russia; foundation of Hapoel Hazair Party.

1906 Foundation of Poalei Zion Party.

1908 Opening of the Palestine Office, and of Palestine Development Company.

1908 Foundation of workers' "pre-moshav" Ein Ganim (near Petach Tikva); first collective nucleus (Segera).

1909 Establishment of Hashomer Organization; foundation of Tel Aviv.

1910 Establishment of the first kvutza (Degania).

1911 First Conference of Agricultural Workers in Judaea and Galilee and foundation of Agricultural Workers' Association.

1913 Foundation of Federation of Judaean Colonies.

1914 Foundation of General Zionists (right wing) party (in New York).

1917 The Balfour Declaration; beginning of Hechalutz Movement in Russia.

1918 Conquest of Palestine by the British Army.

1919 Pogroms in the Ukraine; beginning of Third Aliya; first large kvutza experiment (Kineret); foundation of Achdut Haavoda Party.

1919 E. Yaffe's pamphlet on the establishment of moshavim.

1920 Determination of Principles of Jewish Colonization of Palestine (London conference); establishment of Foundation Fund—Keren Hayessod; acquisition of large tracts of land in the Jezreel Valley by the Keren Kayemeth; establishment of Labor Battalion—Gdud Haavoda; founding conference of Jewish Federation of Labor—Histadruth.

1921 Establishment of Kibbutz Ein Harod, and moshavim Nahalal, Kfar Yehezkiel; establishment of experimental agricultural station; foundation of the Hagana defence organization.

1922 Confirmation of the British mandate in Palestine by the League of Nations.

1923 First split in Ein Harod.

1924 Beginning of Fourth Aliya.

1926 First conference of moshavim and "organizations."

1927 Establishment of Hakibbutz Haartzi of Hashomer Hatzair movement, and of Hakibbutz Hameuchad.

1929 Arab disturbances in the country.

1930 Beginning of Fifth Aliya; establishment of moshav organization.

1932 The Settlement of the Thousand in new moshavim.

1933 Rise of Hitler to power and immigration from Germany.

1934 Beginning of Youth Aliya.

1936–
39 Arab disturbances.
1936 Beginning of defense settlements.
1938 Commission of the Partition of Palestine.
1939 British Conference in London; publication of "White Paper" (on Jewish immigration to Palestine).
1948 Establishment of the State of Israel.
1949 Beginning of mass immigration and of new immigrants' moshavim.

Notes

Introduction

1. Dubious, that is, without very careful and especially designed study in agricultural economics. This would involve not only the examination of the differential efficiency and productivity of the various factors, but also their weighting as to relative scarcity and importance. Such a study, even to the limited extent that it is possible, was clearly beyond our capacity.

Chapter 1. History of the Jewish Settlement in Palestine: A Survey

1. The reader can find a more comprehensive presentation of the subjects discussed in this survey in A. Ruppin, *Three Decades of Palestine: Speeches and Papers on the Upbuilding of the Jewish National Home* (Jerusalem: Schocken, 1936); A. Bein, *The Return to the Soil: A History of Jewish Settlement in Eretz Israel*, in Hebrew (Jerusalem: Youth and Hechalutz Department of the Zionist Organization, 1952); and Palestine Royal Commission, W. R. Wellesley Peel, chairman (London: H. M. Stationary Office, 1937).

2. *Aliya* means immigration. Thus, First Aliya = first wave of immigration. There were several such waves, as shall be seen later on.

3. Edmond de Rothschild, the Paris banker and philanthropist, who supported many Jewish undertakings and charities.

4. The vineyards were concentrated mainly in the Judaea area.

5. Palestine was part of the province of Syria.

6. The reference is to the Keren Hayessod, which is discussed later.

7. From this period onward changes were effected in its duties. These will be dealt with later.

8. To avoid confusion, it must at once be stressed that the *kvutza* was the prototype from which later different collective variations developed. Most of these new variations were called kibbutzim (themselves of several subtypes) so as to distinguish them from the original form, which became a minority; the generic name of kibbutz is now customarily applied to all the collectives. Some of the distinctions between these offshoots from the same branch are discussed later on, when directly relevant to our theme.

9. Ottoman laws, which remained in force after 1918, following the transfer of suzerainty of Palestine from the Ottoman Empire to the British Empire.

10. During the entire period of its existence, the Zionist movement favored investment of private capital in agriculture and industry. Beginning with the 1920's, investments in agriculture were directed mainly to the citrus-plantation area. Farmers in the older colonies of the Judaea area, who originally received help from Baron de Rothschild, were the first to invest in citrus-growing. Later, the fourth wave of immigrants also invested in citrus-growing. Investments in industry were made on a small scale by the Fourth Aliya, but mainly by the fifth wave of immigrants.

11. The attacks were particularly severe in the years 1921, 1929, and 1936 to 1939.

12. Article 4 of the mandate.

13. This right acquired legal status only in 1927, but was recognized *de facto* in 1920 with the first election of the institutions on the initiative of the Jewish community itself.

14. Some of these institutions, principally the educational ones, also received assistance from the Zionist Organization.

15. Data based on Bein, *op. cit.*, pp. 263–264.

16. From the point of view of size, most of these enterprises were small, with ten workers or less. Indeed, till the twenties, there was only one large plant, the wine cellars, while the rest were workshops. Between 1922 and 1925, however, several bigger plants were

established, notably a cement works and a soap and cosmetics factory, and this expansion in size has continued.

17. The Arab attacks on Jewish life and property, begun in 1936, impelled the mandatory government to propose a territorial division of Palestine between the two communities. True, the proposal was rejected by both communities, but until it became an accomplished fact, the Jews attempted to extend the areas in their possession as much as possible in order to create a *de facto* situation that would enlarge the territory on which they had settled, should the proposal be carried out.

18. The Jewish population of Palestine.

Chapter 2. A Colony: Petach Tikva

1. G. Kressel, *Petach Tikva, The Mother of the Colonies,* in Hebrew (Petach Tikva: Petach Tikva Municipality, 1953), pp. 37–39.

2. *Ibid.*

3. *Ibid.*, p. 78.

4. Y. Yaari and M. Charizman, *The Jubilee Book of Petach Tikva,* in Hebrew (Tel Aviv: The Jubilee Committee, 1929), pp. 9–38.

5. *Ibid.*

6. *Ibid.*

7. S. Yavnieli, "The Rules of Petach Tikva," in *The Book of Zionism,* in Hebrew (Tel Aviv: The Bialik Institute, with Our Publications, 1942), Vol. 2, Book 1, p. 96.

8. J. H. Shechter (ed.), *The Memoirs of Zerach Berendt* (Jerusalem: 1929).

9. From a description by one of the first settlers, in A. Tropa: *Seventy Years of Petach Tikvah,* in Hebrew (Petach Tikva: Petach Tikva Municipality, 1948), p. 32.

10. *Ibid.*, p. 48.

11. *Ibid.*, p. 28.

12. *Ibid.*, p. 29.

13. Yaari and Charizman, *op. cit.*, pp. 264–265.

14. Kressel, *op. cit.*, p. 142.

15. *Ibid.*, p. 165.

16. Achad Haam, "The Yishuv and His Patrons," in *Collection* (Tel Aviv: Dvir, 1947), p. 221 n.

17. Klonymus Zeev Wissotzky, *Letters Sent to Important People Concerning the Settlement of Palestine*, in Hebrew (Warsaw: Published privately by the author, 1898).

18. *Ibid.*, p. 299.

19. A. Liebrecht, "The Agriculture of Petach Tikva Since its Beginning," in Hebrew, *Bustanai*, Vol. X, No. 27 (1938), p. 37.

20. J. Rabinovitz, "Petach Tikva," in Hebrew, *Hapoel Hatzair*, Vol. IV, No. 17 (1911), 5.

21. I. Barzilai (Eisenstadt), in Hebrew, "From Eretz Israel," *Hashiloah*, Vol. XIII (1904).

22. Kressel, *op. cit.*, p. 276.

23. J. Rabinowitz, *Hapoel Hatzair*, Vol. IV, No. 18 (1911), 1.

24. J. Aharonowitz, "The Problem of the Workers," in Hebrew *Hapoel Hatzair*, Vol. V, No. 2 (1911), 5.

25. Z. Yoeli and A. S. Shtein, *Conquerors and Builders*, in Hebrew (Petach Tikva: Petach Tikva Labor Council, 1955), p. 102.

26. A description written in 1903, quoted in Yaari and Charizman, *op. cit.*, pp. 264–265.

27. *Ibid.*

28. J. Grazovsky, "From Eretz Israel," in Hebrew, *Hashiloah*, Vol. I (1897), pp. 441–442.

29. J. Rabinowitz, "Petach Tikva as it Really Is," in Hebrew, in Yoeli and Shtein, *op. cit.*, pp. 158–159.

30. It has not, unfortunately, been possible to match the years of the pictures with those of the available land-use records, on which our maps are based; at the same time, a fairly close approximation has been obtained.

Chapter 3. A Kibbutz: Ein Harod

1. Quoted in R. Cohen: *Foundations of the Kibbutz Economy*, in Hebrew (Ein Harod: Hakibbutz Hameuchad, 1946), p. 168 and pp. 174–175.

2. S. Lavee, *My Story in Ein Harod*, in Hebrew (Tel Aviv: Am Oved, 1947).

3. Quoted in Cohen, *op. cit.*

4. Lavee, *op. cit.*, pp. 19–20.

5. *Ibid.*, p. 21.

6. I. Tabenkin, in *Hakibbutz Hameuchad Anthology*, in Hebrew (Tel Aviv), No. 8, 1932.

7. For more details, see Y. Erez (ed.), *The Labor Battalion Anthology*, in Hebrew (Tel Aviv: Mizpe, 1932).

8. See Chapter 6.

9. For more general information, see Erez, *op. cit.*

10. After the Ein Harod conflict, the Battalion severed all its relations with the Histadruth and drifted further to the left. Finally, some of its leaders became completely discouraged with the chances of realizing their radical socialist ideas in Palestine, and reimmigrated, with some of their followers, to Soviet Russia; the unit as a whole was later disbanded.

11. See Chapter 6.

12. Lavee, *op. cit.*, pp. 316–319.

13. Z. Gilad (ed.), *Forty Years of Ein Harod*, in Hebrew (Ein Harod: Hakibbutz Hameuchad, 1962).

14. See *The Situation of Ein Harod—Trends and Developments*, in Hebrew (Ein Harod: Hakibbutz Hameuchad, 1940), pp. 9 and 24.

15. *Ibid.*

16. *Ein Harod Annals*, in Hebrew, No. 13 (1946), pp. 40–42.

17. *The Situation of Ein Harod*, p. 39.

18. See Chapter 6.

19. *The Situation of Ein Harod*, pp. 46–47.

20. Significantly enough because of a split in party allegiance, mainly over questions of foreign policy (see Chapter 6).

21. A term invented by the German sociologist Schmallenbach and first applied to the kibbutz by Yonina Talmon-Garber, in "Social Structure and Family Size," *Human Relations*, Vol. XII, No. 2 (1959), pp. 120–146.

22. *The Situation of Ein Harod.*

23. B. Habas (ed.), *Youth Aliya Book*, in Hebrew (Jerusalem: Department of Youth Immigration, The Jewish Agency, 1941).

24. For a discussion of problems of identity among traditional Youth Aliya children, see C. Frankenstein, "Youth Aliya and the Education of Immigrants," in *Between Past and Future* (Jerusalem: Szold Foundation, 1953).

25. See for instance M. E. Spiro, *Venture in Utopia* (Cambridge, Mass.: Harvard University Press, 1956); Y. Talmon-Garber, "The Family in Collective Settlements," *Transactions of Third World Congress of Sociology*, Vol. IV (1956); "The Family in a Revolutionary Movement," in M. Nimkoff, ed., *Comparative Family Systems* (Boston: Houghton, Mifflin, 1965); and "Social Change and Family Structure," *International Social Science Journal*, Vol. XIV, No. 3 (1963), pp. 468–487.

26. Though later much clearer sex-role differentiation emerged, with the women concentrated chiefly in services. See Y. Talmon-Garber, "Sex-Role Differentiation in an Egalitarian Society" in T. G. Lasswell, J. Burma, S. Aronson (eds.), *Life in Society* (Chicago: Scott, Forsman, 1966).

27. Here, too, Ein Harod, due to its large and diversified social and economic structure, anticipated a process which in many collectives appeared only later. See, for example, E. Rosenfeld, "Social Stratification in a 'Classless' Society," *American Sociological Review*, Vol. XVI, No. 6 (1951), pp. 766–774; R. D. Schwartz, "Functional Alternatives to Inequality," *American Sociological Review*, Vol. XX, No. 4 (1955), pp. 424–430; and Y. Talmon-Garber, "Differentiation in Collective Settlements," *Scripta Hierosolymitana*, Vol. III (Jerusalem: Hebrew University, 1956), pp. 153–178.

28. It is interesting to note that this problem, common to most collectives, has been in the forefront of their minds all the time and has been one of the avowed reasons for their recent experimentation with regional economic enterprises. Such enterprises, owned jointly by several settlements and managed as a unit, are not only efficient economically, but also allow labor to be hired outside the individual village and not directly by it. Moreover, in most cases, each collective owns shares in proportion to the manpower it itself supplies; the enterprise usually has an outside partner (like the Regional Council or the Histadruth) to which the hired labor is apportioned. See, for example, *Regional Cooperation in Israel* (Rehovot: Settlement Study Center, 1965).

29. On this phenomenon today, see Y. Peres, "General Assembly of Members in the Kibbutz," in Hebrew, *Ovnaim*, Vol. III (1962), pp. 76–107.

30. See A. Etzioni, *The Organizational Structure of the Kibbutz* (Ph.D. dissertation, Berkeley: University of California, 1958).

31. Lavee, *op. cit.*

32. See, however R. Gruneberg, "Education in the Kibbutz," in N. Bentwich (ed.), *A New Way of Life: The Collective Settlements of Israel* (London: Shindler and Colomb, 1949), pp. 81–98; S. Golan, "Social Foundations of Kibbutz Education," *Israel Horizons*, January, 1953, pp. 11–13; and S. Golan, *The Theory of Collective Education* (Johannesburg: Hashomer Hatzair, 1952).

33. *Ein Harod Annals.*

34. There is also a religious kibbutz movement, Hakibbutz Hadati.

35. Thus Passover, commemorating the Exodus from Egypt, was given a strong tinge of national liberation and revival, at the expense of such ritual as unleavened bread, etc.; similarly, Shavuoth emphasized the agricultural symbol of bringing first fruits, at the expense of the element of the giving of the Tora (The Law Books).

36. This brief analysis of recent trends in the kibbutzim is based on the works of Y. Talmon-Garber, quoted above. For the significance of these processes in the context of social change in Israel in general, see for example: S. N. Eisenstadt, *Absorption of Immigrants* (London: Routledge & Kegan Paul, 1954); J. Matras, *Social Change in Israel* (Chicago: Aldine, 1965); S. N. Eisenstadt, *Israeli Society* (New York: Basic Books, 1968).

Chapter 4. A Moshav: Nahalal

1. T. Hertzl, *Altneuland* trans. L. Levensohn (New York: Bloch, 1941).

2. J. Wilkansky, "Group of Settlers," *Chapters of Hapoel Hatzair*, in Hebrew, Vol. VIII, Part II, 20.

3. S. Dayan, *Man and His Land: Twenty-five Years of Nahalal*, in Hebrew (Tel Aviv: Massada and Tenuat Hamoshavim, 1948), p. 16.

4. *Ibid.*

5. Cited by S. Dayan in *Nahalalim*, in Hebrew (Tel Aviv: Massada and Tenuat Hamoshavim, 1961), pp. 49–50.

6. J. Wilkansky, "The Village and the City in the Activity of Keren Kayemeth," in Hebrew, *Chapters of Hapoel Hatzair*, Vol. VIII, Part II, 82.

7. E. Jaffe, "Toward the Establishment of Moshavei Ovdim," *ibid.*, 191–192.

8. The numbers given include workers and their families.

9. Wilkansky, "Group of settlers."

10. In fact, the holding envisaged for a moshav unit and adopted as a basis in subsequent villages was half the size: 50 dunams (12.5 acres).

11. The increase by 1936 represents temporary workers, refugees then coming from Nazi Germany.

12. *Bulletin of Nahalal,* in Hebrew, No. 77 (September, 1958).

13. S. Dayan, "Facing the Coming Events," *Nahalal—the 20th Anniversary*, in Hebrew (Tel Aviv: The Secretariat of the Moshavim Organization, 1941), p. 3.

14. *Bulletin of Nahalal,* in Hebrew, No. 37 (November, 1952).

15. Dayan, *Man and His Land.*

16. "The Agricultural Conference of Hapoel Hatzair, Kineret 1929," in Hebrew, *Tlamim* (November–December, 1936), p. 10.

17. *Bulletin of Nahalal,* in Hebrew, No. 81 (March, 1959).

18. *Ibid.,* No. 67 (June, 1957).

19. *Ibid.,* No. 102 (November, 1961).

Chapter 5. Growth of Colonies

1. The association included the grape-growers of Rishon Le Zion, Rehovot, Petach Tikva, Ness Ziona, Gedera, Zikhron Yaakov and Hadera.

2. Quoted by permission of the late A. Hurwitz, from his private collection.

3. The Arabs used to bring denunciations about the Jewish community to the Turkish authorities.

4. The distribution of seats was as follows: workers' lists—111; Oriental Jewish communities—72; religious parties—64; citizens' (right-wing) lists, including the farmers' federation—67.

5. D. Yudelevitz (ed.), *The Book of Rishon Le Zion,* (1882–1941), (Hevrat Carmel Mizrachi: Rishon Le Zion, 1941), p. 327.

6. M. Smilansky, *Selected Chapters in the History of the Yishuv,* in Hebrew (Tel Aviv: Dvir, 1939–1947), p. 47.

7. E. Hadani: *Fifty Years of Settlement in Lower Galilee,* in

Hebrew (Tel Aviv: Massada and Farmers' Association of Lower Galilee, 1955) p. 321.

8. *Ibid.*, p. 353.

9. The franc was an accepted unit of payment in the Levant.

10. From Yavniel—40 members; Mescha (Kfar Tabor)—37; Sedgera—33; Beit-Gan—16; Menahemiya—16; Kinnereth—13; Poriya—16; Migdal—18; Beitaniya—10; Mitzpe—10; Workers' Federation in the Galilee—6; Kevutzat Degania—10; Hashomer—6; total —230 members.

11. These changes are described in the discussion of the following period.

12. At the founding session the following colonies took part: Gedera, Ekron, Rehovot, Beer Yaacov, Ness Ziona, Rishon Le Zion, Petach Tikva, Kfar Saba, Ein Hai, Hadera.

13. For description, see next chapter.

14. From a letter sent by the new organization's central committee to the Beer Yaacov farmers, in which the latter are requested to give financial support to consolidate the organization.

15. From the private papers of the late A. Horwitz.

16. In this organization the following were members: all the farmers of Raanana, Magdiel, Ramat Gan, Bnei Brak, Herzlia, Ein Ganim, Ramat Hasharon, Hadar, Pardess Hanna, and some of the farmers of Petach Tikva, Ness Ziona, and from colonies in lower Galilee.

17. The first moshavot, it will be recalled, were supported by Hovevei Zion, Baron Rothschild, and J.C.A.

18. Within the present context we cannot, of course, analyze this surely fascinating difference between the two farmer groups. Such an analysis would in any case require a special study of a Galilee colony, similar to that undertaken in Petach Tikva. Risking an educated guess, however, the basic factor seems to have been the introduction in the southern colonies of a successful cash crop, which later continuously drove to constant specialization. Such a cash crop—in the form of the grape and wine branch—had been initiated, it will be recalled, by the Baron's administration. Although the settlers themselves later introduced and developed a new citrus monoculture, the crucial transition had been facilitated for them. Not so in the Galilee, where the settlers remained for a long time poor, subsistence farmers. This was so, it seems, not just because the

Baron's innovation did not originally extend to the distant north; after all, the diffusion was only a matter of a hundred miles. However, the Galilee was characterized, as has been mentioned, by poorer land and by a scarcity of water, which for a long time prevented any intensification and which was bound to settle and reinforce the farmer in his static, limited, and extensive cultivation. The fact is that the development of the Galilee collectives and cooperatives also trailed behind, and their take-off was late and difficult. This, however, was made up by their special characteristics and their national organizations, so different from those of the farmers in question. To what extent the objective difficulties and the communicative distance were also underlined by smaller and less active élites, is a question that must be answered empirically.

19. For an examination of both the Hagana and of Achdut Haavoda, see the following chapter.

20. Not so, by the way, in the World Zionist Organization, in which nonlabor parties for a long time had a commanding influence.

21. Through the pressure of other right-wing elements in the assembly, 30 corporate representatives of communities, municipalities, and local councils were admitted in 1938 to the assembly, and among them there were 9 representatives of colonies from the association. These 30 representatives were an addition to the 71 original members of the assembly, as follows: workers' parties—30; right-wing parties—18; religious parties—5; ethnic lists—18.

22. The immigation to Palestine was controlled by the mandatory government, which developed an increasingly restrictive policy in this area. It thus limited the yearly quota of immigrant visas—or "certificates"—and also decided on their division among different categories, such as capitalists, youths, and pioneers. However, the actual allocation of entry permits under this arrangement was entrusted to the Jewish Agency. Additional aspects of this issue will unfold as the story goes on.

Chapter 6. Growth of the Kibbutz Movement: Hakibbutz Hameuchad

1. That is, subject to general legal provisions and chiefly the later restrictions on purchase of lands and location of villages.

2. A fairly large part of the resources had to be allocated to the orthodox religious sector of the Jewish community, which was in dire need of financial and welfare aid. This caused a great deal of dissatisfaction, as this sector was considered especially undeserving. Furthermore, very little control was exercised over the manner in which this aid was distributed.

3. On these parties, see later.

4. In 1924 there were ten thousand members in the Histadruth.

5. See details in M. Braslavsky, *The History of the Jewish Labor Movement,* in Hebrew (Tel Aviv: Hakibbutz Hameuchad, 1955–1959), Vol. I, 111–114, 121–124.

6. In the elections to the first Histadruth convention, Achdut Haavoda received 48 percent of the votes and the Hapoel Hatzair 31 percent.

7. See Z. Rosenstein, *The History of the Labor Movement in Eretz Israel,* in Hebrew (Tel Aviv: Am Oved, 1956), p. 446. For reference to the organizational structure of the Histadruth see *The Histadruth, 1945–1948,* in Hebrew (Tel Aviv: General Federation of Labor, 1949), p. 13.

8. Organized in 1919 as a result of the union between the Poalei Zion and groups of nonparty workers.

9. The party second in size, it will be recalled, and its major opponent until they united in 1930.

10. Of which the best known now are the Cereal Growers', the Cattle Breeders', the Poultry, the Vegetable Growers', the Citrus Growers', and the Sheep Breeders' Associations.

11. A. Manor, *Essence of the Histadruth,* in Hebrew (Dalia, 1957), p. 142.

12. *Ibid.*, p. 143.

13. All members of the Histadruth are also members of Chevrat Haovdim, and the Histadruth convention serves as the general meeting of the Chevrat Haovdim.

14. A detailed list can be found in P. Naftali, *Chevrat Haovdim,* in Hebrew (Tel Aviv: General Federation of Labor, 1946), p. 102.

15. L. Ben-Ari, *Kontres,* in Hebrew, Vol. VII, No. 123 (1923), 4–5.

16. The settlements participated in providing the share capital of Hamashbir in accordance with the gross business they did with it. The kibbutz bought machinery, agricultural tools, fuel, fertilizer,

etc. from Hamashbir. It should be noted that settlements also made purchases from private firms, either because they were unable to obtain the items wanted from Hamashbir or because the credit extended by Hamashbir was inadequate. Purchases from private merchants increased as their economic situation improved and with the growth of the working capital of the settlements. Hamashbir reacted by improving the flow of supplies and providing better services, terms of sale, and credits. See also Apter, "The Directions of Activities of Hamashbir," *Hapoel Hatzair*, in Hebrew, Vol. XXXI No. 2 (1930), 9–11; and Y. Guelfat, "Hamashbir and Its Aim," *Hapoel Hatzair*, Vol. XVIII No. 32 (1925), 8–9.

17. As explained above, all these institutions were outside the labor sector.

18. B. Katzenelson, one of those who conceived the idea of the bank, and a leader of the labor movement. See *Ketavim* (collection of publications) in Hebrew (Tel Aviv: Mapai, 1946), Vol. I, 347.

19. There were three categories of shares:

1) Ordinary shares, which entitled the holders to vote at a general meeting, to participate in the election of the management, and to receive dividends.

2) Preferred shares, earning a fixed, regular dividend, entitled the holders to participate in the election of the management, but they had no vote at the meeting of shareholders.

3) 100 founders' shares. The holders did not share in the profits, but had the right to participate in the election of the management at a general meeting and could cast a number of votes equal to all the other shareholders together.

20. A total of £P 38,000. See Naftali, *op. cit.*, p. 82.

21. *Solel Bone*, a building and contracting subsidiary, which operated in the urban sector; see M. Braslavsky, *op. cit.*, Vol. I, 259.

22. Indeed, as late as 1936, the moshavot comprised 72 per cent of the Jewish rural population and farmed 64 per cent of the cultivated area (including 90 per cent of the citrus), while in 1923, Petach Tikva alone had more settlers than all the workers' settlements together; and the value of the produce of three large colonies (Rishon Letzion, Rehovoth, and Hedera) was greater than that of all the collective and the cooperative villages.

23. *Hakibbutz Hameuchad Anthology*, in Hebrew (Tel Aviv: Hakibbutz Hameuchad, 1932), p. 117.

24. *Ibid.*

25. Z. Rosenstein in M. Basok (ed.), *The Book of the Hechalutz,* in Hebrew, (Jerusalem: The Jewish Agency, 1940), p. 505.

26. A. Tzisling in Basok, *op. cit.,* p. 209.

27. The reference is to the Fourth Aliya.

28. One thousand adults among 22,500 members of the Histadruth.

29. Braslavsky, *op. cit.,* Vol. I, pp. 381–382.

30. The kibbutz, it was contended, "was a social unit of people banded together for the purpose of forming a social cultural cooperative that, by virtue of its very existence, imposed positive and negative duties on its members, whereas a party was mainly a political organization aiming to achieve power and to rule. A party organization took no interest in the manner of life its members led and did not demand of them personally to live in accordance with certain ideals; it merely appealed to people to support its political platform. The kibbutz represented the complete way of life for each member, while the party was only part of it, the political part." See Z. Shafer, *The Growth of a Society,* in Hebrew (Tel Aviv: Am Oved, 1960), pp. 101–102.

31. *Hakibbutz Hameuchad Anthology,* p. 104.

32. The outstanding representatives of these groups were Dr. Chaim Weizman, President of the World Zionist Organization (and later the first President of Israel) and Dr. Arthur Ruppin, director of its Colonization Department.

33. Braslavsky, *op. cit.,* Vol. IV, 96–99.

34. Hakibbutz Hameuchad, *The Resolutions of Conventions and Council Meetings of the Kibbutz,* in Hebrew (Ein Harod: 1955), pp. 3–6.

35. *Ibid.*

36. *Ibid.,* pp. 1–3.

37. In reply to these accusations, which—as we shall see—were voiced also by the moshavim, sharp words were heard from the leaders of the kibbutz at the Yagur Council in December, 1928. Among other things, Y. Tabenkin said: "In the demand not to inculcate the specific aims of the kibbutz ideals for Eretz Israel, I see perfidiousness on the part of the people who have abandoned this way of life. To describe the kibbutz only as an instrument is not correct, as it has a definite content, and is one of the basic aims of our education."

38. Y. Tabenkin, "The Criteria of Immigration," in Basok, *op.*

cit., p. 420. E. Dobkin, "An Answer to an Accusation," in Basok, *op. cit.* p. 509.

39. It will be recalled that the British mandate adopted a highly restrictive immgration policy, limiting the yearly quota on the one hand, and dividing it into categories—such as owners of capital, pioneers, and youths—on the other. Within these restrictions, however, the distribution of permits was entrusted to the Zionist Organization Executive.

40. The Revisionist movement was an exception. It was an extreme right-wing party, which at a later date organized the immigration of its members from Europe to Palestine in the same manner as the Histadruth.

41. E. Dobkin, "The Immigration, Its Composition and Its Absorption," in Basok, *op. cit.*, pp. 412–413.

42. Z. Rosenstein, "A Summary," in Basok, *op. cit.*, pp. 505–506.

43. Naan was founded in 1932 by the *Hanoar Haoved* (Working Youth). Bet-Hashita was established in 1934 by the *Machanot Haolim* (the youth movement in the schools).

44. Hakibbutz Hameuchad, pp. 42–43.

45. Mapai came into being as a result of the merger between the Achdut Haavoda and Hapoel Hatzair parties.

46. A Tzisling, *Mebifnim* (From Inside), in Hebrew (January, 1938), p. 66.

47. For a more detailed description see, Bein, *op. cit.*, pp. 337–338, 423–438, 442–445, 448–450, 455–6, and 475–489.

48. During this period, 230,000 immigrants arrived in the country, characterized by a relatively high proportion of professionals and a considerable amount of capital.

49. In many cases, the workers were specifically called upon to take direct action and decide on policy themselves, and then to compel the executive to adopt their policy. See *Hakibbutz Hameuchad Anthology*, pp. 153–162.

50. During this period, plugot were also assigned to secure key positions in two large industrial undertakings: The Dead Sea Potash Works and the Electric Corporation, which were formed in that period and in a short time became the two largest industrial enterprises in the country.

51. As far back as 1925, 4 members of Kibbutz Ein Harod, out of a total of 158, served as delegates to the first elected assembly. In

1931, they were two out of 70, and in 1944, eleven out of 177 delegates—despite the fact that they accounted for only 2 per cent of the Jewish community.

52. It should not be completely ignored that in those years some of the central personalities of Mapai began, as indicated above, to function as members of the Zionist Executive and thus, of course, made room for a new cadre of leaders to rise to the top.

53. The Underground Self-Defense Organization of the Jewish national institutions.

54. Palmach were commando units of the Self-Defense Organization, consisting of volunteers serving on a permanent basis for a period of two years and based usually on the kibbutzim, established in 1941. For additional information see B–Z Dinur (ed), *History of the Hagana,* in Hebrew (2 vols; Tel Aviv: The Zionist Library and Maarchot, 1954); Z. Ghilad and M. Meghed (eds.), *The Book of Palmach,* in Hebrew (Tel Aviv: Hakibbutz Hameuchad, 1954); and J. Bauer, *Diplomacy and Underground in Zionism 1939–1945,* in Hebrew (Tel Aviv: Sifriat Poalim, 1963).

55. J. Bauer, *op. cit.*, pp. 160, 161, 178, 251.

56. *Ibid.*, pp. 174–175.

57. Naftali, *op. cit.*, pp. 28–29.

58. In the middle of the 1930's, the kibbutzim of Hakibbutz Hameuchad constituted 50 per cent of the agricultural settlements connected with Tnuva.

59. It should be recalled though, that in 1944, despite rapid agricultural expansion, the Jewish community was far from being self-sufficient in the provision of its food supplies. In fact, only 47 per cent of the requirements was supplied by Jewish producers, 47 per cent was imported, and 6 per cent was supplied by Arabs. See Naftali, *op. cit.*, p. 33.

60. Naftali, op. cit., p. 84.

61. *Ibid.*, pp. 90–91.

62. The extent to which the kibbutz still possesses these characteristics, and is mobilized for special pioneering tasks, is exemplified by the recently settled region of Lachish, in which the collectives are situated chiefly along the borders and the cooperative villages occupy the inner area, which is safer from marauding (as well as more fertile).

63. This urban periphery was not really small; at its strongest, it

commanded more voters then Hakibbutz Hameuchad had members.

64. An excerpt from the movement constitution, quoted in *Hakibbutz Hameuchad: Resolutions of Hakibbutz Councils, From the Founding Council in 1927 up to the Daphna Council in 1955*, in Hebrew (Tel Aviv: Chevra, 1962).

65. The extreme left wing of the Zionist labor movement.

Chapter 7. Growth of the Moshav Movement: Tnuat Hamoshavim

1. The nuclei of prospective moshav settlers were called organizations (*irgunim*), to distinguish them from the collective companies (*plugot*).

2. A. Assaf, *Moshavei Ovdim in Israel,* in Hebrew (Ayanot and Tenuat Hamoshavim, 1954), p. 53.

3. *Ibid.*, p. 54.

4. Settled on land in 1927, also in the Jezreel Valley.

5. Assaf, *op. cit.*, pp. 51–52.

6. *Ibid.*

7. A. Ben Arieh, "Problems of the Growth of the New Settlements," *Tlamim* (December, 1934), p. 17.

8. Resolutions of the Third Conference of the Moshavim Movement, *Tlamim* (August, 1935), p. 94.

9. That is, until the new Moshav Bill, of much more recent standing. Thus, the decision on collective contracts was reinforced again and again during subsequent meetings, but the relationship between the Jewish National Fund and the settler has nevertheless remained on an individual basis.

10. "Current Problems," *Tlamim* (August, 1941), p. 2.

11. See S. N. Eisenstadt, *Studies in Social Structure* (Jerusalem: Research Seminar in Sociology, the Hebrew University, 1950) and *From Generation to Generation—Age Groups and Social Structure* (Glencoe, Ill: Free Press, 1956).

12. A. Wilkomitz, "The Moshavim Fund," *Tlamim* (March, 1943), p. 16.

13. J. Shapira, "Our Activities (a Report)," *Tlamim* (July, 1937), p. 26.

14. See, for instance, G. M. Meier (ed.), *Leading Issues in Devel-*

opment Economics: Selected Materials and Commentary (New York: Oxford University Press, 1964).

15. Assaf, *op. cit.*, p. 71.

16. *Ibid.*, p. 72.

17. For detailed analysis, see chiefly S. N. Eisenstadt, *Absorption of Immigrants* (London: Routledge & Kegan Paul, 1954).

18. Assaf, *op. cit.*

19. *Ibid.*

20. Table 27 shows the planned investment program for each farm unit, 70 per cent of which was to be by direct public financing (based on 1958–59 price levels in Israeli pounds).

21. Each such local team was composed of three resident instructors: a vocational-agricultural instructor, a social instructor (who was also the village administrator in the initial stage), and a woman social instructor.

22. A detailed examination of these problems and of the solutions attempted is beyond the scope of the present discussion, and the reader is referred to J. Ben-David (ed.), *Agricultural Planning and Village Community in Israel* (Paris: Unesco, 1964), pp. 13–44.

However, the crucial turning points, introduced over the period of 1953–56, were: (a) a more realistic evaluation of the resources at the disposal of agricultural development; (b) a more realistic evaluation of market potential, and the transition from agricultural autarchy to integration in world markets; and (c) a consequent change in farm planning, which now became differentiated in relation to the above mentioned factors, instead of the hitherto predominant mixed-dairy farm; (d) a radical change in the *pattern* of farm-development based on the assumption that development should be phased over a prolonged period and that the problems of each stage should be anticipated. Responsibility for farm management was gradually transferred from the settlement agencies and the instructor team to the settlers themselves. A new term, the "managed farm," was introduced to describe the immigrants' settlement in its early stages; (e) regional settlement planning, in which production and services are integrated on the regional level.

23. For a vivid description of these problems see A. Weingrod, *Reluctant Pioneers: Village Development in Israel* (Ithaca, N. Y.: Cornell University Press, 1966), pp. 121–142.

24. D. Weintraub *et al.*, *Social Change and Rural Development:*

Table 27. The planned investment for each farm unit

Type of investment	Total investment value (for 80 farms)	Investment per farm unit
Farm house		
2½ rooms (45 square meters), water installations, shower, shutters, outside water-closet	492,000	6,150
Communal buildings		
2 classrooms	17,000	
Creche and kindergarten	13,000	
Assembly hall and club	14,000	
Grocery	12,000	
Clinic	8,000	
Armoury	7,000	
Office	6,000	
Kindergarten kitchen	12,000	
Synagogue	11,000	
Ritual bath	7,000	
Total	107,000	1,338
Farm buildings		
Multipurpose shed (24 square meters) with concrete floor		1,750
Open farm shed (24 square meters) with tin roof, 2 walls		850
Chicken run and equipment for 300 chickens (15 square meters)		780
Stable (30 square meters)		750
Poultry batteries for 100 layers		250
Total	350,040	4,380
Communal farm buildings		
General warehouse-storeroom (210 square meters)	18,000	
Sheep dip (at village center)	4,000	
Grading Station for vegetables (at village center)	4,600	
Technical supervision of farm buildings and houses (6 per cent of total value)	58,640	
Total	85,340	1,067
Irrigation equipment		
Central and lateral piping (permanent) for 6 acres of irrigated land, including orchards, sprinklers, and 4 meters aluminium per acre		4,320
Water meter		160
Total investment in irrigation equipment	358,400	4,480

Table 27. (Cont.)

Type of investment	Total investment value (for 80 farms)	Investment per farm unit
Equipment		
Half-share in auxiliary implements for draught animals		460
Poultry equipment		100
Total	44,800	560
Stock		
100 pullets or cocks and chicks		400
8 head of sheep or 2 beef cows		800
Half-share in traction animal or one work cow		450
Total	132,000	1,650
Settler's share in heavy equipment (at rural center)	93,200	1,240
Orchards		
Planting and maintenance of 1-acre orchard	192,000	2,400
Electrical installations (including cost of linking to main national grid)	30,000	375
Roads		
To storehouse, center of village, etc.	24,000	300
Deep ploughing		
Cost of ploughing 7.25 acres of irrigable land	23,000	288
Revolving capital		
For 5 acres at I£ 400 per acre	160,000	2,000
Water rights, etc.	80,000	1,000
Total investment in equipment and stock	2,178,140	27,228
Planning, instruction and administration costs (15 per cent of I£ 27,200)	326,721	4,084
Total investment	2,504,861	31,300

Source: Lakhish Region planning section.

Immigrant Villages in Israel (Manchester, Eng.: Manchester University Press, 1969 and Jerusalem: Israel Universities Press, 1969).

Chapter 8. Comparative Analysis

1. In this historical context we refer only to broad cultural patterns, capable of being reconstructed; we obviously, though unfortunately, could not refer to motivational factors, associated with patterns of socialization, the status of the family of origin, and other

elements such as those that have been so interestingly analyzed in great detail by D. C. McClelland, in respect to entrepreneurship. See *The Achieving Society* (Princeton, N.J.: Van Nostrand, 1961).

2. This refers especially to groups of settlers who arrived at Petach Tikva from Eastern Europe.

3. We have called this an "almost complete alternative" and the "eliminating of most of its fundamental characteristics" since a complete negation is not to be inferred here. Rather, a selective choice was to be made from the sources of traditional Judaism accompanied by a new structuring of the economic, political, and social components of Jewish society.

4. For further general discussion of this problem, based on empirical data, see A. Etzioni, *Research Report on Division of Labor in the Kibbutz*, in Hebrew (mimeo; Department of Sociology, Hebrew University, 1956).

5. Though, of course, the basis, the scope, and the intensity of the *Gemeinschaft* differ, that of Petach Tikva having been formed essentially after the image of the traditional-religious village in the Diaspora (the *Shtetl*).

6. This does not mean, of course, that the settlers of the moshava have no values transcending economic ones, but simply that their agrarian institutions are divorced or separated from these values.

7. We are told, though, that a study of relative capital productivity in the kibbutz and the moshav is now being prepared by the Department of Agricultural Economics of the Hebrew University.

8. K. W. Deutsch, "Social Mobilization and Political Development," *American Political Science Review*, Vol. LV, No. 3 (1961), 493–515.

9. It will be recalled that, at a later stage (1930), both these parties united and the two settlement movements found themselves under the cover of one party, Mapai.

10. Data based on: (1) colonies: D. Gurevitz and A. Gerz, *Jewish Agricultural Settlement in Palestine* (Jerusalem: Department of Statistics of the Jewish Agency for Palestine, 1938), pp. 42–43 and Table 22; (2) Hakibbutz Hameuchad: A. Israeli, *Milestones—Hakibbutz Hameuchad in Dates and Numbers*, in Hebrew (Ein Harod: Hakibbutz Hameuchad, 1955); (3) moshavim: *The Histadruth in the Thirties*, in Hebrew (Tel Aviv: General Federation of Labor, 1951), p. 33.

Chapter 9. General Conclusions

1. Of course, mobilization of international capital, public and private, is now open to most developing countries, through loans, investment, and even outside grants. The uniqueness of external capitalization in the Yishuv (and later in Israel) lay, however, in that its source and character freed it from any type of foreign control or domination, the fear of which, whether justified or not, is certainly a major impediment to the acceptance of external capital by most new nations. See, for example, G. M. Meier, *Leading Issues in Development Economics: Selected Materials and Commentary* (New York: Oxford University Press, 1964), and B. Hoselitz, "Patterns of Economic Growth," in *Sociological Aspects of Economic Growth* (Glencoe, Ill.: Free Press, 1960), ch. 4.

2. That is, in Hoselitz's terms, the development was expansionist and not intrinsic, able to colonize and build up new and empty spaces rather than having to intensify and reform existing structures. See Hoselitz, *op. cit.*

3. That is, as compared with most other colonial societies. See I. Wallenstein (ed.), *Social Change—The Colonial Situation* (New York: John Wiley, 1965).

4. For an analysis of the Yishuv under the mandate see S. N. Eisenstadt, "The Sociological Structure of the Jewish Community in Palestine," *Jewish Social Studies,* Vol. X (1948), 3–18.

5. Albert Hirschman, *The Strategy of Economic Development* (New Haven: Yale University Press, 1958).

6. We say "seems," as Rostow does not say so explicitly, and it can only be inferred. See W. W. Rostow, *The Process of Economic Growth* (New York: Norton, 1952).

7. Chiefly S. N. Eisenstadt, "Breakdowns in Modernization," *Economic Development and Cultural Change,* Vol. XII, No. 4 (1964).

8. See D. Weintraub *et al., Social Change and Rural Development: Immigrant Villages in Israel.*

9. The dangers inherent in one dominant form are exemplified, of course, by the *ejidos* of Mexico, which for a long time constituted the backbone of rural development and agrarian reform and which were recently found to have become largely stagnant and a drag

upon development. It seems indeed that only the quick action of new entrepreneurs, introducing novel forms (such as young enterprises of farmers and urban capital), and the flexibility of the government in recognizing the change, prevented the rural sector from sliding back in its development.

10. See, for example, S. N. Eisenstadt, *Modernization: Protest and Change* (New York: Prentice-Hall, 1966), Chs. 3 and 6.

11. See Edward Shils, "Centre and Periphery," in *The Logic of Personal Knowledge* (London: Routledge and Paul, 1961).

12. Karl Polanyi, *The Great Transformation: Origins of Our Time* (London: 1947). See also, Moshe Lissak, "Modernization and Role Expansion of the Military in Developing Countries—A Comparative Analysis," *Comparative Studies in Society and History*, Vol. XIV, No. 3 (1967).

13. S. N. Eisenstadt, "Political Modernization: Some Comparative Notes," *International Journal of Comparative Sociology*, Vol. V, No. 1 (1964), 4.

14. For an illuminating analysis of modern and traditional political systems, see C. S. Whitaker, Jr., "A Dysrhythmic Process of Political Change," *World Politics*, Vol. XIX, No. 2 (1967), 190–202.

15. See S. P. Huntington, "Political Development and Political Decay," *World Politics*, Vol. XVII, No. 3 (1965), 387.

16. S. N. Eisenstadt, *The Social Structure of Israel* (New York: Basic Books, 1967).

17. See G. A. Almond and J. C. Coleman (eds.), *The Politics of the Developing Areas* (Princeton, N.J.: Princeton University Press, 1960), pp. 27–28.

18. See M. J. Levy, Jr., "Patterns of Modernization and Political Development," *The Annals of American Academy of Political and Social Science*, Vol. CCCLVIII (1965), p. 35.

19. For example, in 1873 two thirds of the government income came from agriculture. See B. P. Dore, *Land Reform in Japan* (New York: Oxford University Press), p. 75.

20. *Ibid.*, p. 14.

Selected Bibliography

Rural Settlement and Social Structure in Palestine and Israel

Bein, A. *The Return to the Soil: A History of the Jewish Settlement in Israel.* Jerusalem: Youth and Hechalutz Department of the Zionist Organization, 1952.

Ben-David, J. "The Kibbutz and the Moshav," in J. Ben-David (ed.), *Agricultural Planning and Village Community in Israel.* Paris: Arid Zone Research Series, UNESCO, 1964.

Eisenstadt, S. N. *Studies in Social Structure.* Jerusalem, Israel: Research Seminar in Sociology, The Hebrew University, 1950.

——. *Israeli Society.* New York: Basic Books, 1968.

Etzioni, A. "The Functional Differentiation of Elites in the Kibbutz," *American Journal of Sociology*, Vol. LXIV, No. 5 (1959), pp. 476–487.

Lissak, M. "Patterns of Change in Ideology and Class Structure in Israel," *The Jewish Journal of Sociology*, Vol. VII, No. 1 (1965), pp. 46–61.

Rokach, A. "The Development of Agriculture in Palestine and Israel," in J. Ben-David (ed.), *op. cit.*

Rosenfeld, E. "Social Stratification in a 'Classless' Society," *American Sociological Review*, Vol. XVI, No. 6 (1951), pp. 766–774.

Schwartz, R. D. "Functional Alternatives to Inequality," *American Sociological Review*, Vol. XX, No. 4 (1955), pp. 424–430.

Spiro, M. E. *Venture in Utopia.* Cambridge, Mass.: Harvard University Press, 1956.

Talmon-Garber, Y. *The Family in Collective Settlements*, Transactions of Third World Congress of Sociology, Vol. IV, 1956, pp. 116–126.

——. "Social Change and Family Structure," *International Social Science Journal*, Vol. XIV, No. 3 (1962), pp. 468–487.

Talmon-Garber, Y. "The Family in a Revolutionary Movement," in N. Nimkoff (ed.), *Comparative Family Systems*. New York: Houghton, Mifflin, 1965, pp. 259–286.

——. "Sex-Role Differentiation in an Egalitarian Society," in T. E. Lasswell, J. Burma, S. Aronson (eds.), *Life in Society*. Chicago, Ill.: Scott, Forsman, 1966, pp. 144–145.

Weintraub, D. "A Study of New Farmers," *Sociologia Ruralis*, Vol. IV, No. 1 (1964), pp. 3–49.

——, and Bernstein, F. "Social Structure and Modernization: A Comparative Study of Two Villages, *American Journal of Sociology*, Vol. LXXI, No. 5 (1966), pp. 509–521.

——, and Lissak, M. "The Moshav and the Absorption of Immigrants," "Physical and Material Conditions in the New Moshav," "Social Integration and Change," in J. Ben-David (ed.), *op. cit.*, pp. 95–159.

Modernization and Rural Development

Almond, G. A., and Coleman, J. C. (eds.). *The Politics of Developing Areas*. Princeton, N.J.: Princeton University Press, 1960.

Apter, D. *The Politics of Modernization*. Chicago, Ill.: University of Chicago Press, 1965.

Befu, H. "The Political Relations of the Village to the State," *World Politics*, Vol. XIX, No. 4 (1967), pp. 601–620.

Belshaw, C. S. "Social Structure and Cultural Values as Related to Economic Growth," *International Social Science Journal*, Vol. XVI, No. 2 (1964), pp. 217–218.

Deutsch, K. W. "Social Mobilization and Political Development," *American Political Science Review*, Vol. LV, No. 3 (1961), pp. 493–514.

Dore, R. P. *Land Reform in Japan*. New York: Oxford University Press, 1959.

Eisenstadt, S. N. "Political Modernization: Some Comparative Notes," *International Journal of Comparative Sociology*, Vol. V, No. 1 (1964), pp. 3–24.

——. "Breakdowns of Modernization," *Economic Development and Cultural Change*, Vol. XII, No. 4 (1964), pp. 345–367.

——. *Modernization: Protest and Change*. New York: Prentice-Hall, 1966.

Hirschman, Albert. *The Strategy of Economic Development.* New Haven, Conn.: Yale University Press, 1958.

Hoselitz, B. *Sociological Aspects of Economic Growth.* Glencoe, Ill.: Free Press, 1960.

Huntington, S. P. "Political Development and Political Decay," *World Politics,* Vol. XVII, No. 3 (1965), pp. 386–430.

Levy, M. J., Jr. "Patterns (Structures) of Modernization and Political Development," *Annals of the American Academy of Political and Social Science,* Vol. CCCLVIII (1965), pp. 29–40.

Lissak, M. "Modernization and Role Expansion of the Military in Developing Countries—a comparative analysis," *Comparative Studies in Society and History,* Vol. XIV, No. 3 (1967), pp. 233–255.

Meier, G. M. (ed.). *Leading Issues in Development Economics.* New York: Oxford University Press, 1964.

Nash, M. "Some Social and Cultural Aspects of Economic Development," *Economic Development and Cultural Change,* Vol. VII, No. 2 (1959), pp. 137–157.

Nettl, J. P. *Political Mobilization: A Sociological Analysis of Methods and Concepts.* New York: Basic Books, 1967.

Polanyi, Karl. *The Great Transformation: Origins of Our Time.* New York: Farrar and Rinehart, 1944.

Rostow, W. W. *The Process of Economic Growth.* New York: Norton, 1952.

Shils, Edward. "Centre and Periphery," in *The Logic of Personal Knowledge.* London: Routledge and Kegan Paul, 1961.

Wallenstein, I. (ed.). *Social Change—the Colonial Situation.* New York: John Wiley, 1965.

Whitaker, C. S., Jr. "A Dysrythmic Process of Political Change," *World Politics,* Vol. XIX, No. 2 (1967), pp. 190–216.

Index